PHYSICS
for Class XII

PHYSICS
for Class XII

Nikhat Khan

OXFORD
UNIVERSITY PRESS

OXFORD
UNIVERSITY PRESS

Great Clarendon Street, Oxford OX2 6DP

Oxford University Press is a department of the University of Oxford.
It furthers the University's objective of excellence in research, scholarship,
and education by publishing worldwide in

Oxford New York

Auckland Cape Town Dar es Salaam Hong Kong Karachi
Kuala Lumpur Madrid Melbourne Mexico City Nairobi
New Delhi Shanghai Taipei Toronto

with offices in

Argentina Austria Brazil Chile Czech Republic France Greece
Guatemala Hungary Italy Japan Poland Portugal Singapore
South Korea Switzerland Turkey Ukraine Vietnam

ISBN 978-0-19-547338-4

Typeset in Times & Helvetica
Printed in Pakistan by
New Sketch Graphics, Karachi.
Published by
Ameena Saiyid, Oxford University Press
No. 38, Sector 15, Korangi Industrial Area, PO Box 8214
Karachi-74900, Pakistan.

Contents

Foreword

Physics for Class XII is primarily meant for students at the higher secondary school level, but will also be useful for BSc students, as it is an introductory one-year textbook for students majoring in Biology, Chemistry, Mathematics, Health Sciences and other disciplines.

The main objective of this textbook is to strengthen a student's understanding of the principles and concepts of Physics in a clear and logical manner. It is well written—simple and clear—with solved examples and numerous problems at the end of each chapter, to help students develop problem solving abilities. To facilitate optimal comprehension, the author explains the concepts of Electricity and Magnetism, Electronics and Modern Physics through illustrations, diagrams and interesting applications.

Dr Nikhat Khan has more than 20 years of teaching Physics at college level to her credit. Her colleagues and students recognize her dedication and effective teaching strategies. Most of her students enjoy a difficult subject like Physics because she makes it easy due through her communication skills and the excellent teacher-student relationship she establishes.

Professor Dr Mira Phailbus
Principal
Kinnaird College for Women
Lahore

1 Electrostatics

OBJECTIVES

After studying this chapter, a student should be able:
- to understand and differentiate between the two types of charges and how bodies gain/lose charge.
- to understand how charge is conserved.
- to understand interaction and forces between charged bodies.
- to understand that electric field intensity varies with distance.
- to understand lines of electric force and their meaning.
- to understand electric potential and absolute potential and the relation between electric field and potential.
- to understand electric flux and Gauss's Law and its applications.
- to understand electron volt – a small energy unit.
- to understand that smallest charge is the charge on an electron/proton.
- to understand about capacitors, capacitance and equivalent capacitance for capacitances connected in series and parallel.
- to be able to derive the equation for capacitance of a parallel plate capacitor.
- to understand that energy is stored in the electric field of a capacitor.
- to understand that capacitors take time in charging and discharging.
- to understand why polarization of the dielectric reduces the electric field between the plates of a capacitor.

1.1 Introduction

Life today would be inconceivable without electricity and magnetism. Imagine living without light, fans, refrigerators, motors and dynamos, computers and cars!

Electricity and magnetism are manifestations of the electromagnetic force.

This chapter encompasses the origin of electricity, types of charges, electric fields, electric flux, electric potential and capacitors. Students are already familiar with the basic electromagnetism concepts. In this textbook, more advanced topics as well new topics will be presented.

1.2 Electric Charge

Electric charge is a part of matter and matter is made up of atoms. An atom has in its nucleus positive charge (protons) and neutral particles (neutrons). Negative charge (electrons) revolve around the nucleus.

There are only **two kinds of charges: positive and negative.**

Using pith balls it can be demonstrated that like charges repel and unlike charges attract each other. If two positively or negatively charged pith balls are brought close to each other they will repel. If one is positively charged and the other negatively charged they will attract each other.

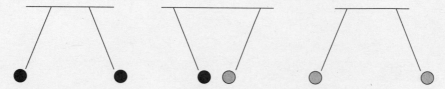

Fig. 1.1: Like charges repel and unlike charges attract.

In Figure 1.1 two positive charges, protons, shown as grey, and the two electrons (black) would repel and positive and negative charges would attract each other. **The symbol for charge is q or Q.**

The charge on an electron and proton is the **smallest, elementary charge** that can exist independently in nature. Charge is an intrinsic property of protons and electrons. It is equal to $\pm 1.6 \times 10^{-19}$ C. The magnitude of charge on an electron is equal to that on a proton i.e. 1.6×10^{-19} C. Charges greater than this are integral multiples of this smallest charge. **Charge is thus quantized.**

$$\text{Charge } Q = ne \qquad\qquad 1.1$$

where e = 1.6×10^{-19} C and n is an integer (n = 1, 2, 3,....)

Example: How many electrons would carry a charge of 1C?

$$1.6 \times 10{-}19 \text{ C} = 1 \text{ electron}$$
$$1 \text{ C} = \frac{1}{1.6 \times 10^{-19}}$$
$$= 6.25 \times 1018 \text{ electrons}$$

Thus we see that 6.25×10^{18} electrons carry a charge of 1 coulomb.

A large number of electrons (6.25×10^{18} electrons) will have to leave a body to carry a net charge of 1C. A charge of one coulomb is a large charge, so charges are normally measured in **milli-coulombs mC (10^{-3} C), micro-coulombs μC (10^{-6} C) and pico-coulombs pC/$\mu\mu$C (10^{-12}C).**

Table 1.1 shows the masses and charges of the three fundamental particles.

Table 1.1: Masses and Charges of Elementary Particles

Particles	Masses (kg)	Charge (C)
Proton	1.673×10^{-27}	$+ 1.6 \times 10^{-19}$
Electron	9.109×10^{-31}	$- 1.6 \times 10^{-19}$
Neutron	1.675×10^{-27}	0

Atoms are neutral. Material bodies that we encounter are neutral because they have the same number of protons and electrons. They neutralize each other. **Charge is conserved.**

1.2.1 Conductors and Insulators

Matter is made up of molecules and atoms. In most materials electrons are tightly bound to the nuclei of atoms and cannot leave the atom. In some elements, the electrons are not so tightly bound and can be removed from the atom.

The materials in which the electrons are tightly bound are **insulators (dielectrics).** Materials in which the electrons are not so tightly bound and can move about in the body of the substance are called **conductors.** Metals are good conductors while glass, rubbers, plastics are insulators. Chapter 6 presents a more detailed discussion on insulators and conductors.

1.2.2 Separating charges by rubbing

When we rub two bodies together heat is generated due to friction. The surface electrons gain enough energy to leave the body making it positively charged (it now has more positive charges and less number of electrons) and the other body becomes negatively charged because it gains electrons.

When a glass rod is rubbed with a piece of silk, electrons from the glass rod move to silk making the silk negatively charged while the glass rod is now positively charged. When you comb your hair on a dry day, transfer of electrons occurs and the hair flies. Why?

1.2.3 Air Filters

Air filters are an important application of this separation of charges. These are devices used to clean air of particulate matter like dust, pollen, particles of soot, etc. The polluted air is first passed through a positively charged mesh. The dust particles get positively charged and are then passed through a negatively charged grid (Figure 1.2).

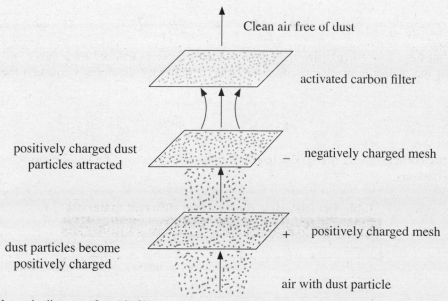

Fig. 1.2: Schematic diagram of an Air filter.

The particles are attracted to this grid and collect on it. It has to be washed after a few days. The air devoid of the dust particles is next passed through an activated carbon filter which absorbs any foul odours before being released into the room.

1.3 Coulomb's Law

Coulomb's Law is related to the force between charges. It was empirically determined and first published in the 18th century by Charles Coulomb. It is an **Inverse Square Law Force** like the gravitational force. Coulomb's Law gives the magnitude of the force. The force acts along the line joining the two charges. **The law states that the magnitude of the force between two charges is directly proportional to the product of the charges and inversely to the square of the distance between them.**

$$F \quad \alpha \quad \frac{q_1 q_2}{r^2} \qquad\qquad 1.2$$

$$F \quad = \quad \frac{k\, q_1 q_2}{r^2} \qquad\qquad 1.3$$

The constant of proportionality k is an electrical quantity that depends upon the medium (air, oil, vacuum, etc) in which the charges are located and the system of units used. In the SI system of units

$$k \quad = \quad \frac{1}{4\pi\varepsilon_o} \quad = \quad 8.99\times10^9 \,\text{Nm}^2/\text{C}^2 \qquad\qquad 1.4$$

where ε_o **is the permittivity of free space and equal to 8.85 x10^{-12} C^2/(N.m^2).**

In solving problems, **k is taken as 9×10^9 Nm^2/C^2**

$$\text{Therefore } F \quad = \quad \frac{1}{4\pi\varepsilon_o}\frac{q_1 q_2}{r^2} \qquad\qquad 1.5$$

When the medium is not free space then the Coulomb's force is reduced by a factor, ε_r, which is a constant for the dielectric medium. ε_r is the relative permittivity. Coulomb's equation then modifies to:

$$F \quad = \quad \frac{1}{4\pi\varepsilon_o\varepsilon_r}\frac{q_1 q_2}{r^2} \qquad\qquad 1.6$$

The dielectric constant for air is 1.00054. In calculations it is taken as 1. Table 1.2 gives some important dielectric constants.

Table 1.2: Dielectric constants for different materials

Material	Relative permittivity ε_r
Vacuum	1
Air	1.00054
Teflon	2.1
Mica	4.5 – 8
Water	80
Rubber	5.1
Paraffined paper	2.0

The Coulomb's force is the force between two point charges. Charges are considered as point charges as the distance is much greater between the charges as compared to the circumference of the sphere carrying the charge.

The Coulomb's force is attractive or repulsive depending on whether the two charges are the unlike or like respectively (Figure 1.3). **It is measured in newtons (N) in the SI system.**

Fig. 1.3: Repulsive and attractive electrostatic forces.

Force is a vector quantity. The direction of the force is given by the unit vector $\mathbf{r}_{12}$ where $\mathbf{r}_{12}$ is directed from charge 1 to charge 2 and $\mathbf{r}_{21}$ from charge 2 to charge1. Therefore $\mathbf{F}_{12}$ is the force on charge 1 due to charge 2 and $\mathbf{F}_{21}$ is the force on charge 2 due to charge 1.

$$\mathbf{F}_{12} = \frac{1}{4\pi\varepsilon_o\varepsilon_r} \frac{q_1 q_2}{r^2} \hat{r}_{12}$$

$$\text{and} \quad \mathbf{F}_{21} = \frac{1}{4\pi\varepsilon_o\varepsilon_r} \frac{q_1 q_2}{r^2} \hat{r}_{21} \qquad 1.7$$

The magnitude of the above forces is equal.

As $\mathbf{r}_{21}$ is opposite to $\mathbf{r}_{12}$ it can be seen that $\mathbf{F}_{12}$ is also opposite to $\mathbf{F}_{21}$

$$\text{In terms of } \mathbf{r}_{12} \quad \mathbf{F}_{21} = \frac{1}{4\pi\varepsilon_o\varepsilon_r} \frac{q_1 q_2}{r^2} (-\hat{r}_{12})$$

$$\text{and so} \quad \mathbf{F}_{12} = -\mathbf{F}_{21} \qquad 1.8$$

If the force on a charged body is due to more than one charged body, first the forces on it are to be found for each individual charge and then the net force on it is the vector sum of these forces. The following examples will make this clear.

Example: Two like charges of $4\mu C$ and $7\mu C$ are placed 1.4 m apart. Find the magnitude and direction of the electrostatic force on $7\mu C$ charge.

$q_1 = 4\mu C$ $\qquad\qquad\qquad\qquad\qquad q_2 = 7\mu C$

1.4m

As the charges are like charges the force will be repulsive

$$F_{coulomb} = \frac{k\,q_1 q_2}{r^2}$$

$$= \frac{9 \times 10^9 \times 4 \times 10^{-6}C \times 7 \times 10^{-6}C}{(1.4)^2\,m^2}$$

$$= 0.128 \text{ N in the direction as shown}$$

Example: Three positively charged particles $5\mu C$, $10\mu C$ and $3\mu C$ are placed as shown. Find the force on the $10\mu C$ charge due to the other charges.

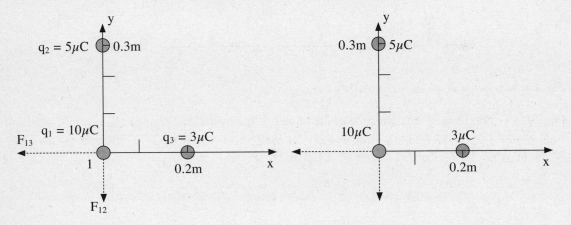

Since all charges are positive the direction of the force on $10\mu C$ due to each of the others is as shown.

The force on q_1 due to q_2

$$F_{12} = \frac{kq_1 q_2}{r_{12}^2} = \frac{9 \times 10^9 \times (10 \times 5)\,10^{-12}}{(0.3)^2} = 3N$$

The force on q_1 due to q_3

$$F_{13} = \frac{kq_1 q_2}{r_{13}^2} = \frac{9 \times 10^9 \times (10 \times 3)\,10^{-12}}{(0.2)^2} = 6.75N$$

The forces F_{12} and F_{13} are shown as dotted lines.

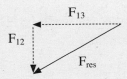

Applying head and tail rule gives the resultant force, since the forces are perpendicular.

$$F_{res} = \sqrt{[(F_{12})^2 + (F_{13})^2]}$$

$$= \sqrt{[(3N)^2 + (6.75N)^2]}$$

$$= 7.39 \text{ N}$$

The direction is given by

$$\tan\theta = \frac{F_{12}}{F_{13}} = \frac{3N}{6.75N} = 0.444$$

$$\theta = 23.9°$$

If the forces were not perpendicular then the sum of the x-components of the forces would be equal to the x-component of the resultant and the sum of the y-component of the forces would be equal to the y-component of the resultant

$$F_{RX} = F_{1x} + F_{2x} \qquad\qquad 1.9$$

$$\mathbf{F_{RX}} = \mathbf{F_1 \cos\theta_1} + \mathbf{F_2 \cos\theta_2} \qquad\qquad 1.10$$

$$\text{and } F_{RY} = F_{1y} + F_{2y} \qquad\qquad 1.11$$

$$\mathbf{F_{RX}} = \mathbf{F_1 \sin\theta_1} + \mathbf{F_2 \sin\theta_2} \qquad\qquad 1.12$$

The resultant force $\qquad \mathbf{F_{resultant}} = \sqrt{[(F_{RX})^2 + (F_{RY})^2]} \qquad\qquad 1.13$

and direction of the resultant force

$$\tan\theta = \frac{F_{RY}}{F_{RX}} \qquad\qquad 1.14$$

Test yourself:
Find the force on particle 1 due to particles 2 and 3 as shown in the figure.
($q = 10^{-6}$ C)

```
    1              2                                    3
    •← 20 cm →• •←————— 50 cm —————→•
   +q             +q                                  −3q
```

1.4 Fields of Force and Electric Intensity

When a charged body is placed in a certain region, it alters the area around itself and we say an electric field has been created. Electric field can be due to one or many charges. When another charged body is placed in this field it experiences a force, the same electrostatic force of section 1.3.

The Electric Intensity **E** is the measure of the strength of the electric field. **It is defined as the force a unit positive charge would experience when placed in the field.**

Therefore mathematically the electric field E is given by

$$E = \frac{F}{q} \qquad\qquad 1.15$$

$$= \frac{k\,q'q\hat{r}}{qr^2} \qquad \textbf{as } F = \frac{k\,q'\,\hat{r}}{r^2}$$

$$= \frac{k\,q'\,\hat{r}}{r^2} \qquad\qquad 1.16$$

where **E** is the electric intensity, q' is the source charge producing the field, r is the distance from q' to q where the field intensity is being computed and **r** the unit vector showing direction.

It can be shown that the Coulomb's force **F** in terms of the electric intensity is given by

$$F = qE \qquad\qquad 1.17$$

The electric intensity **E** has the same direction as the Coulomb's force **F** as can be seen in the above equation.

The principle of superposition can be applied to electric intensity **E**. The electric intensity at any point due to other point charges is the vector sum of the intensities at that point due to the individual charges.

The unit of electric intensity is N/C.

The Electric Intensity **E** is a field quantity and it is a very important quantity. It explains why a charge placed in an electric field immediately, with no time gap, experiences a force.

It is now accepted that a charge has a force field surrounding it. The intensity of this field decreases as distance from the charge increases. As soon as another charge is placed in the field, its field interacts with the field in which it is placed. A resulting force is produced immediately due to this interaction and the charge experiences a force. A positive charge experiences a force in the direction of the field whilst a negative charge experiences a force in the opposite direction (Figure 1.4).

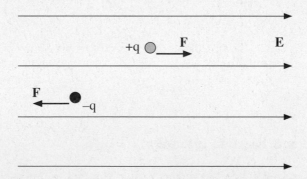

Fig. 1.4: Direction of force experienced by positive and negative charges in an electric field.

Example: The electric field between electrodes of a gas discharge tube used in a neon sign has magnitude 5×10^4 N/C. Find the acceleration of a neon ion of mass 3.3×10^{-26} kg and charge $+ 1.6\times10^{-19}$C. How does this acceleration compare with g?

$$E \quad = \quad 5 \times 10^4 \text{ N/C}$$

$$m \quad = \quad 3.3 \times 10^{-26} \text{ kg}$$

$$q \quad = \quad + 1.6 \times 10^{-19} \text{C}$$

$$F \quad = \quad qE$$

$$= \quad (1.6 \times 10^{-19}\text{C}) \, (5 \times 10^4 \text{ N/C})$$

$$= \quad 8.0 \times 10^{-15} \text{ N}$$

$$\text{Acceleration} \quad a \quad = \quad \frac{F}{m}$$

$$= \quad \frac{8 \times 10^{-15}\text{N}}{3.3 \times 10^{-26}\text{kg}}$$

$$= \quad 2.4 \times 10^{11} \text{ m/s}^2$$

Comparing this acceleration with g

$$\frac{a}{g} \quad = \quad \frac{2.4 \times 10^{11} \text{ m/s}^2}{9.81 \text{ m/s}^2}$$

$$= \quad 244 \times 10^{10} \text{ times!}$$

1.5 Electric Fields Lines

The line of force at a point is the direction in which a unit positive charge would move in an electric field. The **electric field can be depicted by these paths or lines which are called electric field lines.** Lines of force help in visualizing electric fields.

Figure 1.5 shows these field lines. In Figure 1.5a, the field lines due to a +Q charge can be seen and in Figure 1.5b, the field lines due to charges +Q and –Q (a dipole) can be seen.

Fields lines are directed from positive to negative.

Fig. 1.5a   *Fig. 1.5b*

Electric field lines radiate from charges in all directions. An infinite number of field lines can be drawn. Field lines patterns give us important information about the field. If the lines are parallel and equally spaced it means it is a uniform, unidirectional field (Figure 1.6a).

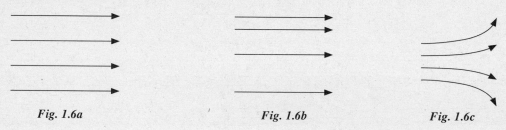

Fig. 1.6a   *Fig. 1.6b*   *Fig. 1.6c*

If the lines are parallel but not equally spaced the field is non-uniform and unidirectional as shown in Figure 1.6b. Figure 1.6c shows a field that is non-uniform and its direction changes as well.

   The number of lines crossing a unit area placed perpendicular to the field is the field intensity. Lines of force are close together in a stronger field and far apart in a weaker field. The number of lines is proportional to the magnitude of the field. A 5 N/C field will have 5 times more lines than a 1 N/C electric field. In Figure 1.6c, it can be seen that the field shows a decreasing intensity from left to right.

   The electric field at any point is tangent to the field line passing through that point.

This is depicted in Figure 1.7a.

Field lines do not cross each other (Figure 1.7b).

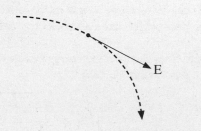

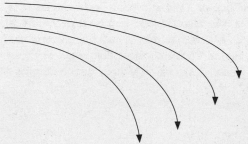

Fig. 1.7a: The electric intensity **E** is tangent to the field line.

Fig. 1.7b: The electric field lines do not cross each other.

1.6 Applications of Electrostatics

There are numerous applications of static electricity. In this section, the Xerox machine or as it is more familiarly called photocopier and inkjet printer shall be discussed.

1.6.1 Photocopier
The photocopier and laser printers work on the separation of charges. A selenium coated aluminium drum is rotated under an electrode. It gets positively charged. Selenium is an insulator but starts conducting when light shines on it. It is a photoconductor.

The drum becomes conducting in places where light falls on it. Therefore when it is illuminated with the image of the document to be copied those regions become conducting and electrons from aluminium neutralize the positive charges of the conducting regions.

The drum is now allowed contact with the toner, a black negatively charged powder. The toner particles stick to the positively charged negatives of the drum.

A positively charged sheet of paper is now rotated on the drum. The negatively charged toner particles are attracted to the paper as it has a higher positive charge than the drum. Thus, an image of the document to be photocopied (xeroxed) is formed on the paper sheet. The paper is then passed through hot rollers so that the toner is attached firmly to the paper fibres and the photocopy is ready (Figure 1.8).

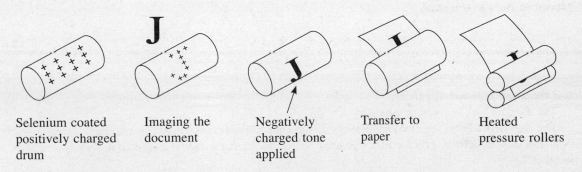

Selenium coated positively charged drum

Imaging the document

Negatively charged tone applied

Transfer to paper

Heated pressure rollers

Fig. 1.8: Photocopier.

1.6.2 An Inkjet Printer
An everyday example of the use of electric force in the computer age is the inkjet printer. The inkjet of the printer moves continuously across the paper ejecting tiny droplets of ink through a nozzle. These minute droplets travel at high speed towards the paper first passing through a charging electrode and then through deflection plates (Figure 1.9).

When the print head moves to regions which are not to be printed then the charging electrodes are put on. The ink drops are charged and the deflection plates deflect these away from the paper. They are thrown away as waste. When printing has to be done the charging electrodes are switched off and the droplets land on the paper.

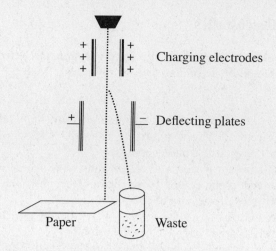

Fig. 1.9: An Inkjet Printer.

1.7 Electric Flux

The electric flux by definition is equal to the dot product of the electric intensity E and area ΔA placed in the electric field.

$$\Phi_e \quad = \quad \mathbf{E}.\,\mathbf{\Delta A} \qquad\qquad 1.18$$

$$= \quad E\,\Delta A\cos\theta \qquad\qquad 1.19$$

θ is the angle between **E** and **ΔA**

The electric flux is a **scalar quantity**.

The flux depends on the electric intensity E, the area ΔA and cosθ where θ is the angle between the field and area directions. [Please note that area is a vector quantity and its direction is perpendicular to its plane].

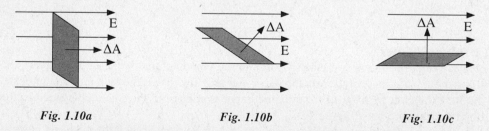

| Fig. 1.10a | Fig. 1.10b | Fig. 1.10c |

Thus, it can be seen that the electric flux will have a maximum value of EA as in Figure 1.10a, E ΔA cosθ in Figure 1.10b and zero in Figure 1.10c. It will range from a minimum of zero where ΔA and **E** are perpendicular to a maximum of EA where **E** and ΔA are parallel.

The flux thus tells us the amount of field through an area perpendicular to the field.

The SI unit for electric flux is N m²/C.

1.8 Flux through a Closed Surface enclosing a Charge

Let us imagine a sphere S that encloses charge q which is placed at the centre of the sphere.

The electric field due to this charge will be radial as shown in Figure 1.11a.

Let us divide the surface area of the sphere into n small areas which are so small that each can be considered as a plane surface. The direction of the electric field at each small area as well as the direction of each small area is the outward drawn normal at that point (Figure 1.11b).

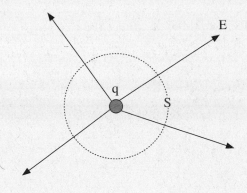

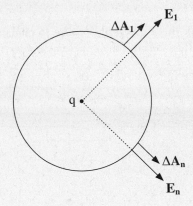

Fig. 1.11a: Radial field through the spherical surface with charge q at its centre.

Fig. 1.11b: Directions of electric intensity **E** and area ΔA at the surface of the sphere S.

The total flux through the sphere is the sum of the fluxes through each small area.

$$\Phi_{sphere} = \Phi_1 + \Phi_2 + \Phi_3 \ldots\ldots\ldots + \Phi_n \qquad 1.20$$

$$= \mathbf{E_1} . \Delta \mathbf{A_1} + \mathbf{E_2} . \Delta\mathbf{A_2} + \mathbf{E_3} . \Delta\mathbf{A_3} + \ldots\ldots + \mathbf{E_n} . \Delta\mathbf{A_n} \qquad 1.21$$

$$= E_1 \Delta A_1 \cos0° + E_2 \Delta A_2 \cos0° + E_3 \Delta A_3 \cos0° + \ldots + E_n \Delta A_n \cos0° \qquad 1.22$$

$$= E \Delta A_1 + E \Delta A_2 + E \Delta A_3 + \ldots\ldots + E \Delta A_n$$

as magnitude of **E** at the sphere's surface is the same in all directions and $\cos0° = 1$

$$\Phi_{sphere} = E [\Delta A_1 + \Delta A_2 + \Delta A_3 + \ldots\ldots\ldots + \Delta A_n]$$

$$= E \Sigma \Delta A_n$$

where $\Sigma \Delta A_n$ is the total surface area of the sphere of radius r and is equal to $4\pi r^2$

Thus, the electric flux through a sphere is

$$\Phi_{\text{sphere}} \quad = \quad E\, 4\pi r^2$$

$$= \quad \frac{kq}{r^2}\, 4\pi r^2 \qquad \text{as} \quad E \quad = \quad \frac{kq}{r^2}$$

$$= \quad \frac{q\, 4\pi r^2}{4\pi\varepsilon_o\, r^2} \qquad \text{as} \quad k \quad = \quad \frac{1}{4\pi\varepsilon_o}$$

$$= \quad q/\varepsilon_o$$

$$\Phi_{\text{sphere}} \quad = \quad q/\varepsilon_o \qquad\qquad\qquad\qquad\qquad 1.23$$

The flux through a sphere due to charge q at its centre is equal to 1/ε_o times the charge q.
The flux through the sphere also goes through the closed surface S as shown in Figure 1.12.

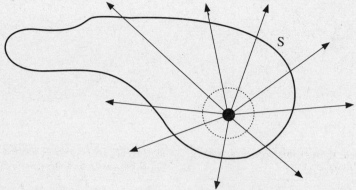

Fig. 1.12: The flux through the sphere also goes through the closed surface S.

1.9 Gauss's Law

Let n charges q_1, q_2, q_3, ...q_n, be enclosed within a surface S.

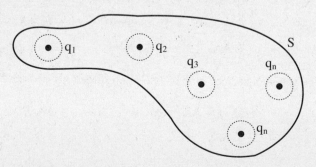

Fig. 1.13: Closed surface S with charges q_1, q_2, q_3.......... q_n.

The flux through the surface S can be found by enclosing each charge by an imaginary sphere (Figure 1.13). Since flux is the dot product of the vector **E** and **ΔA** and is a scalar quantity, it can be added by simple arithmetic. Thus, the flux through surface S is the sum of the fluxes through the n spheres:

$$\Phi_S \quad = \quad \Phi_{\text{sphere 1}} \quad + \quad \Phi_{\text{sphere 2}} \quad + \quad \Phi_{\text{sphere 3}} \quad +\ldots\ldots\ldots+ \quad \Phi_{\text{sphere n}}$$

$$= \quad q_1/\varepsilon_o \quad + \quad q_2/\varepsilon_o \quad + \quad q_3/\varepsilon_o \quad +\ldots\ldots\ldots+ \quad q_n/\varepsilon_o$$

$$= \quad 1/\varepsilon_o\,[q_1 \quad + \quad q_2 \quad + \quad q_3 \quad +\ldots\ldots\ldots+ \quad q_n]$$

where $q_1 + q_2 + q_3 \ldots\ldots+ q_n$ is the total charge Q enclosed in the closed surface S.

Thus

$$\Phi_S \quad = \quad \frac{Q}{\varepsilon_o} \qquad\qquad 1.24$$

where Q is the total charge enclosed by the closed surface. This is also known as **Gauss's Law** and the **closed surface S is known as Gaussian surface**.

Gauss's Law is an important law in electrostatics. **It states that the total flux through a closed surface is $1/\varepsilon_o$ times the charge Q enclosed in the closed surface.**

1.10 Applications of Gauss's Law

Gauss's Law can be applied to determine the electric field intensity due to charges placed on spheres, cylinders, sheets, etc.

The field due to an infinite sheet of charge, the electric field between two sheets and the field inside a closed surface shall be derived using the law. An imaginary and relevant Gaussian surface is selected, Gauss's Law is applied and the electric intensity is determined.

The flux is computed using equation 1.18 and by Gauss's Law (equation 1.24). The two are equated and the electric intensity is so determined. In this section, the electric intensity **E** due to a charged hollow sphere and due to sheets of charge are discussed.

1.10.1 Field inside a uniformly charged hollow sphere

Charge q is placed on a conducting (metallic) sphere. Due to mutual repulsion the charges spread uniformly as far as possible over the surface of the sphere. For this situation a spherical Gaussian surface S of radius $r < R$ is selected (Figure 1.14).

Applying Gauss's Law on the Gaussian surface S

$$\Phi \quad = \quad \frac{\text{Charge enclosed}}{\varepsilon_o}$$

$$= \quad 0 \qquad\qquad 1.25$$

as there is no charge inside S

But flux is also equal to

$$\Phi \quad = \quad \ddot{E}\,\Delta A$$

$$= \quad E\,4\pi\,r^2 \qquad\qquad 1.26$$

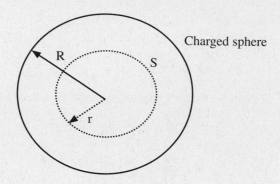

Fig. 1.14: Charged sphere with Gaussian surface at r < R.

Equating equations 1.25 and 1.26 gives

$$\Phi \quad = \quad E4\pi r^2 \qquad = \quad 0$$

But area $A \quad = \quad 4\pi r^2 \neq 0$

therefore $E \quad = \quad 0$ 1.27

In the region r < R there is no electric field. The region inside the sphere is shielded from any field. This is true for any charged conducting object of any shape as shown in Figure 1.15.

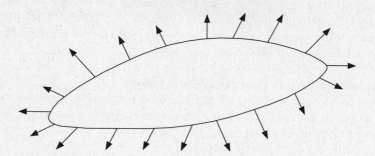

Fig. 1.15: The region inside a hollow surface is shielded from any field.

Charges accumulate more at sharp points where the radius of curvature of the surface is the minimum.

1.10.2 Field due to an infinite sheet of charge

We already know that the electric field is directed from positive to negative, so the electric field due to an infinite sheet of uniform density positive charge would be directed perpendicularly away from the sheet on both sides (Figure 1.17).

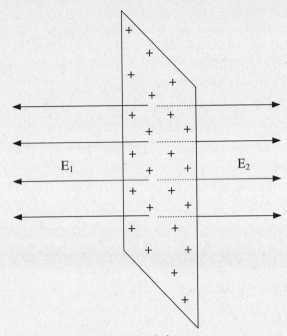

Fig. 1.16: Electric intensity E due to a uniformly charged sheet.

We wish to determine the electric intensity E at point P.

Let us select points P and P′ on both sides of sheet at equal distance from it. To apply Gauss's Law, we imagine a cylindrical Gaussian surface S (dotted line) that encloses both these points as shown in Figure 1.17.

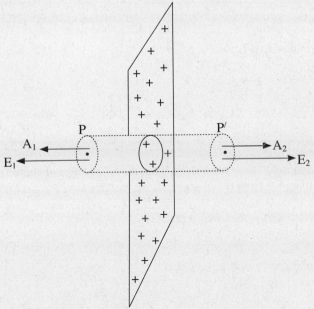

Fig. 1.17: The Gaussian surface S cuts across the uniformly charge sheet through the circle.

We divide the surface S into three areas: two plane surfaces A_1 and A_2 both having directions as outward drawn normal from the plane surfaces and the curved surface A_3 of the cylinder. The curved surface A_3 is further divided into small areas $\Delta A_1, \Delta A_2,........., \Delta A_n$, each with its own direction as seen in the Figure 1.18. Points P and P′ are at the plane surfaces A_1 and A_2 respectively of the closed surface S.

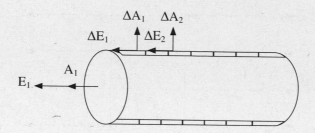

Fig. 1.18: The curved surface is divided into small areas ΔA_n. ΔA_n and **E** are perpendicular to each other at all points on A_3.

The fluxes through the areas A_1, A_2 and A_3 are

$$\Phi_1 = \mathbf{E_1 . A_1} = E_1 A_1 \cos 0° = E_1 A_1 \qquad \qquad 1.28$$

$$\Phi_2 = \mathbf{E_2 . A_2} = E_2 A_2 \cos 0° = E_2 A_2 \qquad \qquad 1.29$$

$$\Phi_3 = \mathbf{\Delta E_1 . \Delta A_1} + \mathbf{\Delta E_2 . \Delta A_2} + ...\mathbf{\Delta E_n . \Delta A_n}$$

$$= \Delta E_1 \Delta A_1 \cos 90° + \Delta E_2 \Delta A_2 \cos 90° + ...\Delta E_n \Delta A_n \cos 90°$$

$$= \Delta E_1 \Delta A_1 + \Delta E_2 \Delta A_2 + ...\Delta E_n \Delta A_n$$

$$= 0 \qquad \qquad 1.30$$

as $\Delta \mathbf{E}$s will be perpendicular to all $\Delta \mathbf{A}$s

Adding 1.28, 1.29 and 1.30

$$\Phi_{total} = \Phi_1 + \Phi_2 + \Phi_3$$

$$= E_1 A_1 + E_2 A_2 + 0$$

Both E_1 and A_1 are magnitudes i.e. scalar quantities. As E_1 and E_2 are the electric fields at P and P′ which are at the same distance from the charged sheet, therefore

$$E_1 = E_2 = E$$

$$\text{Also } A_1 = A_2 = A \qquad \text{(plane surfaces of same cylinder)}$$

$$\Phi_{total} = EA + EA$$

$$= 2EA \qquad \qquad 1.31$$

Applying Gauss's Law which states that the flux is equal to the charge enclosed in a closed surface divided by ε_o.

The charge enclosed by the cylindrical surface is that on the area A (bold circle) of the sheet that is cut by the cylinder.

If   Charge density $= \quad \sigma$

         Charge on area A $= \quad \sigma A$

         Flux Φ $= \quad \dfrac{\sigma A}{\varepsilon_o}$ Gauss' Law   1.32

Equating equations 1.31 and 1.32

$$2EA \quad = \quad \sigma A / \varepsilon_o$$

$$E \quad = \quad \frac{\sigma}{2\varepsilon_o}$$

The direction of E will be along the outward drawn normal given by unit vector **r**

$$\text{Thus } \mathbf{E} \quad = \quad \frac{\sigma}{2\varepsilon_o}\,\hat{r} \qquad\qquad 1.33$$

The equation for **E** gives us important information about the field. As both σ and ε_o are constants, **E** is a constant. It is a uniform field and does not change as distance from the infinite sheet of charge increases.

1.10.3 Electric field due to two uniformly charged sheets

Two sheets of infinite extent are placed parallel to each other as shown in the Figure 1.19.

The electric intensity at point P between the plates is to be determined.

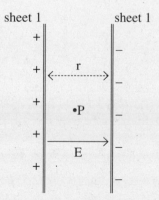

Fig. 1.19: Two charged sheets of infinite extent placed parallel to each other.

To determine the electric intensity at P, we first find $\mathbf{E_1}$ due to positively charged sheet 1 (Figure 1.20) and then $\mathbf{E_2}$ due to negatively charged sheet 2 (Figure 1.21) and then apply the principle of superposition i.e. add $\mathbf{E_1}$ and $\mathbf{E_2}$.

$$E_1 \quad = \quad \frac{\sigma}{2\varepsilon_0}\,\hat{r} \qquad \text{(away from + vely charged sheet)}$$

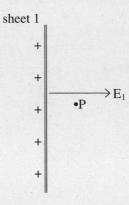

Fig. 1.20: Electric intensity E_1 due to one sheet.

$$\text{and} \quad \mathbf{E_2} \quad = \quad \frac{\sigma}{2\varepsilon_0}\,\hat{r} \qquad \text{(towards the -vely charged sheet)}$$

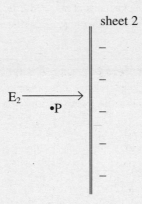

Fig. 1.21: Electric intensity E_2 due to one sheet.

Since both $\mathbf{E_1}$ and $\mathbf{E_2}$ are in the same direction the total electric intensity $\mathbf{E}$ is the sum of $\mathbf{E_1}$ and $\mathbf{E_2}$

$$\mathbf{E} \quad = \quad \mathbf{E_1} \quad + \quad \mathbf{E_2}$$

$$= \quad \frac{\sigma}{2\varepsilon_0}\,\hat{r} \quad + \quad \frac{\sigma}{2\varepsilon_0}\,\hat{r}$$

$$= \quad \frac{\sigma}{\varepsilon_0}\,\hat{r} \qquad\qquad\qquad 1.34$$

The electric intensity between two charged sheets is equal to the charge density σ divided by the ε_0 the permittivity of free space.

Test yourself: If both sheets are positively charged with the same charge density what would be the electric intensity between the sheets?

1.11 Electric Potential

Let us compare the work done in raising a mass m against the gravitational force $\mathbf{F_{grav}}$ to moving a charge q against the electric force $\mathbf{F_{el}}$ (Figure 1.22).

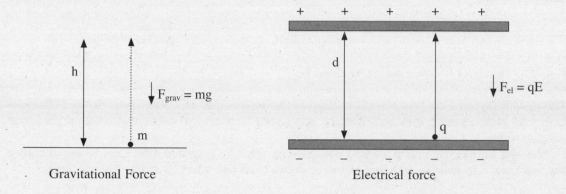

Fig. 1.22: The mass m is raised against the gravitational force and charge q against the electric force.

The work done in both cases:

$$W \;=\; \mathbf{F_{grav}.\,s} \qquad\qquad W \;=\; \mathbf{F_{el}.\,s}$$
$$=\; mg\,h\,\cos 180° \qquad\qquad =\; qE\,d\,\cos 180°$$
$$=\; -mgh \qquad\qquad\qquad =\; -qEd$$

The work done in moving a charge q against the electric force ($\mathbf{F} = q\mathbf{E}$) from A to B through a distance Δs (Figure 1.23) is given by

$$W \;=\; -q\mathbf{E}.\,\Delta s \qquad\qquad\qquad 1.35$$

The electric potential difference ΔV is defined as the work done in moving a unit charge against the electric field through a distance Δs

$$\text{Thus} \qquad \Delta V \;=\; \frac{\Delta W}{q}$$

$$=\; \frac{F.\Delta s}{q}$$

$$=\; \frac{qE.\Delta s}{q}$$

$$=\; \mathbf{E.\,\Delta s}$$

$$= \quad E \, \Delta s \, \cos 180°$$

$$\Delta V \quad = \quad -E \, \Delta s \qquad\qquad 1.36$$

The electric potential is a scalar quantity.

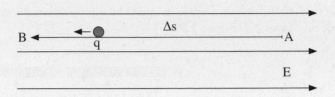

Fig. 1.23: The charge is moved through **Δs** against **E**.

The electric potential ΔV is analogous to change in temperature ΔT in Thermodynamics and change in height Δh in Mechanics. It is the difference in potential between two points that is important not its absolute value.

The unit for electric potential in SI units is volts (V). The potential difference is one volt when the work done in moving a charge of one coulomb from one point to another is one joule.

$$1V \quad = \quad \frac{1J}{1C} \qquad\qquad 1.37$$

1.11.1 Electric Intensity as Potential Gradient
From the equation 1.36 for potential difference it can be seen that the electric intensity **E** is equal to

$$E \quad = \quad -\frac{\Delta V}{\Delta s} \qquad\qquad 1.38$$

In the equation 1.38, **ΔV/Δs is the potential gradient.** The negative sign means that the electric field is opposite to the direction in which the potential increases (Figure 1.24).

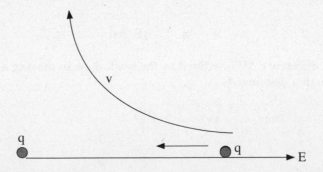

Fig. 1.24: The potential gradient.

The SI unit for the potential gradient and another unit for the Electric Intensity E is V/m.

Example: A uniform electric field E has a magnitude 300 N/C and is directed to the right. A particle with charge 5nC moves along a straight line through 0.3 m.
    a) What is the electric force that acts on the particle?
    b) What is the work done on the particle?
    c) What is the p.d.?

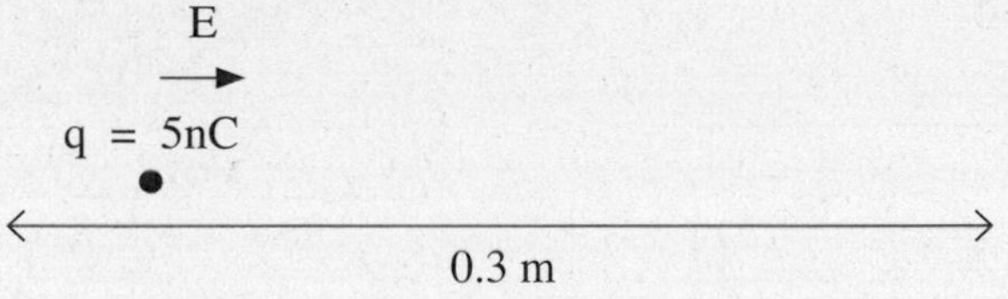

$$E$$

$$q = 5nC$$

$$0.3 \text{ m}$$

$$E \quad = \quad 300 \text{ N/C}$$

$$q \quad = \quad 5 \text{ nC}$$

$$s \quad = \quad 0.3 \text{ m}$$

a)   Force $= \quad q\,E$

$$= \quad 5 \text{ nC} \times 300 \text{ N/C}$$

$$= \quad 1.5 \ \mu N$$

b)   Work $= \quad F\,s$

$$= \quad 1.5 \ \mu N \times 0.4 \text{ m}$$

$$= \quad 0.45 \ \mu \text{ J}$$

c)   p.d. $= \quad \dfrac{W}{q}$

$$= \quad E\,s$$

$$= \quad 300 \text{ N/C} \times 0.3 \text{ m}$$

$$= \quad 90 \text{ V}$$

1.11.2 Electric Potential at a Point due to a Point Charge

The equation $\Delta V = -E\,\Delta s$ for the electric potential is true when the field is a uniform field which generally is not the case. An expression for the potential difference (p.d) when the field is non-uniform is derived.

    Let us assume that an electric field **E** is produced due to a point charge Q placed at the origin of the co-ordinate system (Figure 1.25a). The direction of the electric field will be away from this charge and decreases as distance increases as

$$E \quad = \quad \frac{kQ}{r^2}\,\hat{r}$$

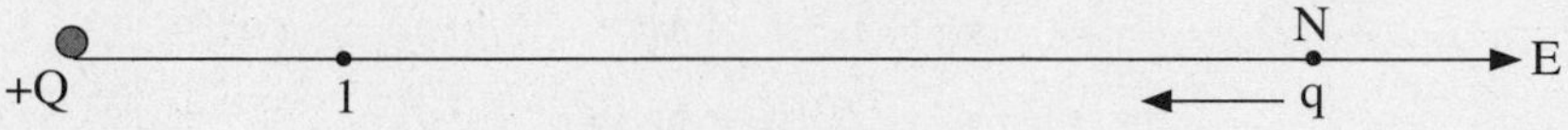

+Q   1   N E
                                   ← q

Fig. 1.25a: Electric intensity due to Q decreases with distance.

Let us move a charge q from a point N to point 1 against **E.** The equation $\Delta V = -E.\Delta s$ is applicable only when E is constant. Therefore, the distance from N to 1 is divided into very small distances each equal to Δr during which distances it is assumed that E is constant (Figure 1.25b).

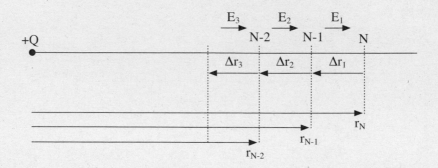

Fig. 1.25b: Points N, N–1, N–2, ... are at distances r_N, r_{N-1}, r_{N-2} ......, r_N. The distances between consecutive points is Δr_1, Δr_2, Δr_3, ...

Let us determine the electric potential form N → N–1. In the distance Δr_1 let the value of electric intensity be E_1

Then

$$E_1 \quad = \quad \frac{kQ}{r^2}$$

To determine the first distance r we will use the average distance i.e. the distance at the centre of r_N and r_{N-1} as the value of r.

$$r \quad = \quad r_{av.}$$

$$= \quad \frac{r_N + r_{N-1}}{2}$$

But $\mathbf{r_{N-1} - r_N = \Delta r_1}$

$$r_{av} \quad = \quad \frac{r_N + (\Delta r_1 + r_N)}{2}$$

$$= \quad \frac{2r_N + \Delta r_1}{2}$$

$$r_{av}^2 \quad = \quad \frac{4r_N^2 + 4r_N\Delta r_1 + \Delta r_1^2}{4}$$

$$= \quad \Delta r_1^2 \text{ is negligible}$$

$$\cong \quad r_N^2 + r_N \Delta r_1$$

$$\cong \quad r_N (r_N + \Delta r_1)$$

$$\cong \quad r_N r_{N-1}$$

$$\text{Thus } E_1 \;=\; \frac{kQ}{r_{n-1}\,r_n}$$

$$\text{But }\quad \Delta V_{N \to N-1} \;=\; \mathbf{E_1}.\,\mathbf{\Delta r_1}$$

$$=\; E_1\,\Delta r_1 \cos 180°$$

$$=\; -E_1\,\Delta r_1$$

$$=\; \frac{-kQ\big[r_{N-1} - r_N\big]}{r_{N-1}\,r_N}$$

$$=\; kQ\left[\frac{1}{r_{N-1}} - \frac{1}{r_N}\right] \qquad\qquad 1.39$$

$$\text{Sim}^{ly}\quad \Delta V_{N-1 \to N-2} \;=\; kQ\left[\frac{1}{r_{N-2}} - \frac{1}{r_{N-1}}\right] \qquad 1.40$$

....

$$\Delta V_{2 \to 1} \;=\; kQ\left[\frac{1}{r_1} - \frac{1}{r_2}\right] \qquad\qquad 1.41$$

To obtain $\Delta V_{N \to 1}$ add RHS of equations 1.39, 1.40,......................, 1.41

$$\mathbf{\Delta V_{N \to 1}} \;=\; kQ\left[\frac{1}{r_1} - \frac{1}{r_N}\right] \qquad\qquad 1.42$$

The potential difference between two points is important.

The **Absolute Potential** is the potential difference between the potential at infinity to a point in the field.

$$\Delta V_{\infty \to r} \;=\; kQ\left[\frac{1}{r} - \frac{1}{\infty}\right]$$

$$V \;=\; kQ\left[\frac{1}{r} - 0\right]$$

$$V \;=\; \frac{kQ}{r}$$

$$V \;=\; \frac{Q}{4\pi\varepsilon_o r} \qquad\qquad 1.43$$

This absolute potential V is the potential difference between the potential at distance r from source charge Q and the potential at infinity which is the reference point and is taken as zero.

An important consequence of this equation $V = kQ/r$ is that whenever the distance is r from source, V will have the same value. If a charge Q is placed at the centre of a spherical surface all points on this surface are equidistant from the centre charge. Thus at all these points V will be the same. Surfaces that have all points on them at the same potential are known as **equipotential surfaces** (Figure 1.26).

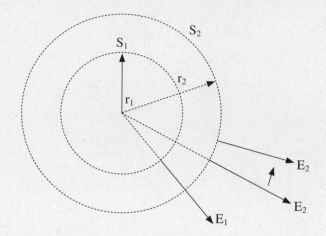

Fig. 1.26: The dotted spheres are equipotential surfaces S_1 and S_2.

The electric field lines are always normal to the equipotential surfaces.

No work is required to move a charge on these surfaces.

1.11.3 Examples of potentials in biological systems

Electric potentials in the body are important for diagnostic purposes. There are small potential differences between various points on the body ranging from 30μV to 500μV. The cell potentials vary as electric impulses move through them. The electrocardiogram (ECG), electro-encephalogram (EEG) and electro-retinogram (ERG) measure the potential differences of cells in the heart, the electrical activity of the brain and the activity in the retina due to a light flash respectively. By comparing the graphs obtained with normal ones, physicians are able to diagnose problems (Figure 1.27).

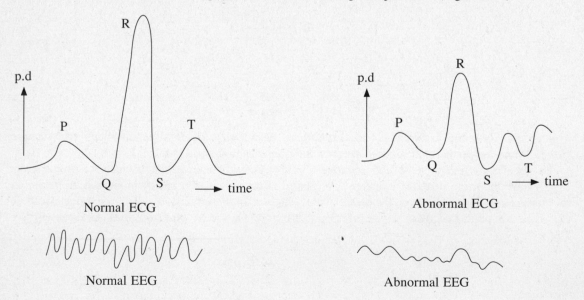

Fig. 1.27: Normal and abnormal ECG and EEG.

1.12 The Electron - Volt (eV)

Like the joule, the electron-volt is a unit of energy. It is widely used in atomic physics.

The electric potential difference is the work done in moving a unit charge from one point to another.

$$\Delta V = \frac{\text{work done}}{\text{charge}}$$

$$\Delta V = W/q$$

Work done is equal to the change in the potential energy ΔU for unit charge

$$\Delta V = \frac{W}{q} = \frac{\Delta U}{q}$$

Thus $\Delta U = q \, \Delta V$ 1.44

If the charge is equal to charge of an electron e and the potential difference is 1 volt (1V) then

$$\Delta U = e \, V \text{ (an electron volt)}$$ 1.45

An eV is the energy gained by an electron that has been accelerated through a potential difference of 1 V.

Putting in the value of e in the equation

$$1 \text{ eV} = (1.6 \times 10^{-19} \text{ C})(1\text{V})$$

$$1\text{eV} = 1.6 \times 10^{-19} \text{ J}$$

An electron-volt (eV) is a small energy unit and is equal to 1.6×10^{-19} J. In nuclear physics, the **MeV (10^6 eV)** and **GeV (10^9 eV)** are commonly used.

Example: What is the velocity of a neutron whose kinetic energy is 60 eV?

[Take mass of a neutron as 1.7×10^{-27} kg].

$$KE = 60 \text{ eV}$$

$$m_n = 1.7 \times 10^{-27} \text{ kg}$$

$$1\text{eV} = 1.6 \times 10^{-19} \text{ J}$$

The kinetic energy in joules is

$$KE = 60 \text{ eV} \times 1.6 \times 10^{-19} \text{ J/eV}$$

$$= 9.6 \times 10^{-18} \text{ J}$$

$$KE = \tfrac{1}{2} mv^2$$

Example: (contd.)

$$v = \sqrt{\frac{[2(KE)]}{m}}$$

$$= \sqrt{\frac{2 \times 9.6 \times 10^{-18}\,J}{1.7 \times 10^{-27}\,kg}}$$

$$= 10.62 \times 10^4 \text{ m/s}$$

1.13 Electric and Gravitational Forces

In this section, a comparison is made between the Electrostatic and Gravitational forces.

Similarities

The two forces are remarkably similar:

Both forces are **inverse square law forces**. They are inversely proportional to the square of the distance between the objects.

Both forces **act along the line joining the centres of the objects.**

Both forces are directly proportional to the product of their intrinsic property — **charges** in Coulomb's Force and **masses** in Gravitational Force.

Both forces have similar equations:

Electric Force $\qquad$ **F** $\quad = \quad \dfrac{kq_1 q_2}{r^2}$

Gravitational Force $\quad$ **F** $\quad = \quad \dfrac{Gm_1 m_2}{r^2}$

Differences

The **electric force can be attractive or repulsive** depending on whether the charges are unlike or like respectively. The **gravitational force is always attractive**.

The gravitational force is **a weak force** as compared to the electrostatic force.

The electrostatic force depends on the **medium**.

1.14 Millikan's Oil Drop Experiment

In section 1.2, it was mentioned that electric charge is quantized, that is any charge is always an integral multiple of the smallest charge i.e. the charge on an electron or proton.

Robert Millikan, an American physicist performed a historic experiment in 1909 to substantiate this. A schematic diagram of Millikan's experiment is shown in Figure 1.28.

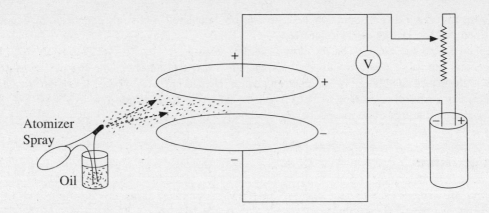

Fig. 1.28: Millikan's Oil Drop Experiment.

Oil drops were passed through a nozzle. As the drops move through the nozzle they acquire charge by friction. They are then subjected to an electric field. The drops experience two forces, one is their weight which acts downwards and is equal to W = mg. The other force on the drop is the electric force (F = qE) which is directed upwards. The electric potential is adjusted so that the forces are equal. The drop is in equilibrium:

Thus $\qquad\qquad\qquad\qquad$ $F_{electric}$ $\quad=\quad$ $F_{gravitational}$ $\qquad\qquad\qquad\qquad$ 1.46

$\qquad\qquad\qquad\qquad\qquad\quad$ qE $\quad=\quad$ mg

The charge on the drop is $\qquad\quad$ q $\quad=\quad$ $\dfrac{mg}{E}$ $\qquad\qquad\qquad\qquad\qquad$ 1.47

As V = Ed $\qquad\qquad\qquad\quad$ q $\quad=\quad$ $\dfrac{mgd}{V}$ $\qquad\qquad\qquad\qquad\qquad$ 1.48

The values of d the distance between the plates and the voltage V can be measured easily.

The mass of the drop can be found by letting the drop fall under the action of the gravitational force only. The circuit is switched off. The drop accelerates downwards. The upward thrust increases and soon the weight becomes equal to the upward force. The drop is in equilibrium and it moves with the terminal velocity v_t. Stokes Law is applicable and so

$\qquad\qquad\qquad\qquad\qquad\quad$ W $\quad=\quad$ F_{Stokes}

$\qquad\qquad\qquad\qquad\qquad\quad$ mg $\quad=\quad$ $6\pi\,\eta\,rv_t$ $\qquad\qquad\qquad\qquad$ 1.49

If the density of the drop oil is ρ, then m = volume × density – 4/3 π r³ ρ

Equation 1.48 becomes

$\qquad\qquad\qquad$ $(4/3\,\pi\,r^3\rho)\,g$ $\quad=\quad$ $6\pi\,\eta\,rv_t$

$\qquad\qquad\qquad\qquad\qquad\quad$ r $\quad=\quad$ $\sqrt{\dfrac{[9\,\eta\,v_t]}{2\,g\,\rho}}$ $\qquad\qquad\qquad\qquad$ 1.50

By putting the value of r in equation 1.48, m can be calculated. Thus, all quantities of equation are determined and the charge easily calculated.

Millikan found the charges on the drops ranging from 1.6×10^{-19} C to 27.2×10^{-19} C. In each measurement the value was 1.6×10^{-19} C multiplied by a whole number. The charges measured were always multiples of the charge on an electron/proton. He did not get a charge less than the charge on an electron or proton i.e. $\pm 1.6 \times 10^{-19}$ C. Thus, he was able to conclude that 1.6×10^{-19} C was the smallest charge that can exist independently in nature.

1.15 Capacitors

Capacitors are devices used to store charge. The RAM in computer chips contains thousands of minute capacitors. Capacitors are very important circuit elements in Electronics. They are responsible for tuning in radios and television. When you tune to FM 100 you are changing the capacitance. Capacitors are also used as filters to remove unwanted frequencies.

A simple capacitor is the parallel-plate capacitor. It consists of two small plates with a dielectric (insulator) between them. When a capacitor is connected to a voltage source, negative charges (electrons) from one plate move to the other till a voltage V equal to that of the source builds up between the plates. Once the capacitor is charged, flow of charges stops (Figure 1.29).

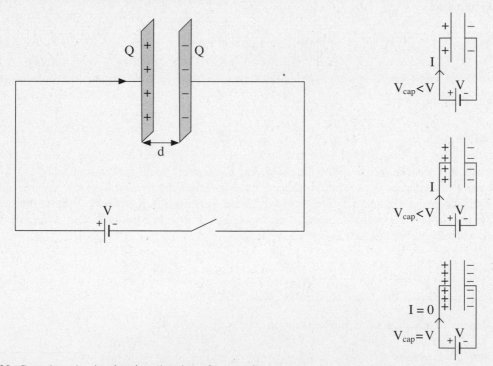

Fig. 1.29: Capacitor circuits showing charging of a capacitor.

Place two voltage sources each of voltage V and the charge stored would be doubled i.e. 2Q. It is found that the charge stored on each plate of the capacitor is proportional to the voltage supplied to them:

$$Q \quad \alpha \quad V$$

$$Q \quad = \quad CV \quad \text{C is the constant of proportionality}$$

$$C \quad = \quad \frac{Q}{V} \qquad\qquad\qquad 1.51$$

C measures the charge storing capacity of the capacitors — how much charge it can store per unit voltage across the plates. It is known as **capacitance** of the capacitor.

The **Capacitance of a capacitor is the amount of charge stored per unit voltage.**

The unit for capacitance is the **farad** named after Faraday.

$$C \quad = \quad \frac{Q}{V}$$

$$= \quad \frac{1 \text{ coulomb}}{1 \text{ volt}}$$

$$= \quad 1 \text{ farad (F)}$$

When a charge of one coulomb is stored on each plate of a capacitor and the voltage across the plates is one volt then the capacitance of the capacitor is one farad.

More commonly used units for capacitance are **millifarads (mF) and microfarads (μF).**

1.16 Capacitance of a parallel plate capacitance

The simplest capacitor is the parallel-plate capacitor which as the name suggests consists of two parallel metal plates with dielectric between them. The capacitance is affected by the area of the plates, the medium between them and how far apart the plates are.

If the electric field between the plates is E, then V = Ed. Putting this value of V in general equation for capacitance

$$C \quad = \quad \frac{Q}{V}$$

$$C \quad = \quad \frac{Q}{Ed}$$

$$C \quad = \quad \frac{Q}{(\sigma/\varepsilon_o)d} \qquad\qquad \text{as } E = \sigma/\varepsilon_o \text{ and } \sigma = Q/A$$

$$C \quad = \quad \frac{Q}{(Q/_{A\varepsilon o})d}$$

$$C \quad = \quad \frac{A\varepsilon_o}{d} \qquad\qquad\qquad 1.52$$

Thus the capacitance is directly proportional to the area of the plates and the permittivity of free space and inversely to the distance between the plates.

If a dielectric medium is placed between the plates of a parallel-plate capacitor, the capacitance is enhanced by a factor of ε_r, the relative permittivity. Now the capacitance is C'

$$C' = \frac{A\varepsilon_o\varepsilon_r}{d} \qquad\qquad 1.53$$

Dividing C' by C

$$\frac{C'}{C} = \varepsilon_r \qquad\qquad 1.54$$

The relative permittivity ε_r is the ratio of the capacitance of the parallel-plate capacitor with the dielectric medium between the plates to that of capacitance of the parallel-plate capacitor with the medium as free space.

Example: A parallel-plate capacitor has plates of area $0.8m^2$ and plate spacing 0.6mm. The dielectric used is Teflon.
  a) What is the capacitance?
  b) If the potential across the plates is 24 V, how much charge is stored on each plate?

$$
\begin{aligned}
\text{Area A} \quad &= \quad 0.8m^2 \\
d \quad &= \quad 0.6mm \quad = \quad 0.0006\ m \\
V \quad &= \quad 24\ V \\
\varepsilon_o \quad &= \quad 8.85 \times 10^{-12}\ C^2/(N.m^2) \\
C \quad &= \quad ? \\
Q \quad &= \quad ?
\end{aligned}
$$

a) For teflon

$$\varepsilon_r = 2.1 \text{ (Table 1.2)}$$

$$
\begin{aligned}
C' &= \frac{A\varepsilon_o\varepsilon_r}{d} \\
&= \frac{0.8m^2 \times 8.85 \times 10^{-12} C^2/(N.m^2) \times 2.1}{0.0006m} \\
&= 2.48 \times 10^{-8}\ F \\
&= 0.025\ \mu F
\end{aligned}
$$

b)

$$
\begin{aligned}
Q &= CV \\
&= (2.48 \times 10^{-8} C/V)\ (24V) \\
&= 0.59 \times 10^{-6}\ C \\
&= 0.59\ \mu C
\end{aligned}
$$

1.16.1 Computer application of the parallel-plate capacitor

An important application of the parallel-plate capacitor is in the familiar computer keyboard. In some type of keys a parallel-plate capacitor is connected.

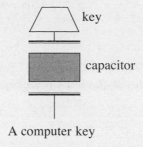

A computer key

As the key is pressed, it pushes the plate of the capacitor. The distance d between the plates of the capacitor decreases. This results in an increase in the capacitance and the circuit of that key is activated.

1.17 Effect of Dielectric

When a dielectric is placed between the capacitor plates, capacitance increases. The capacitance C′ is given by

$$C' \quad = \quad \frac{A\varepsilon_o\varepsilon_r}{d}$$

For a parallel-plate capacitor, it is enhanced by the factor ε_r which is the relative permittivity. For air ε_r is slightly more than 1. In some texts, ε_r is sometimes termed as dielectric constant.

Table 1.2 gives the values of ε_r for different insulating (dielectric) materials.

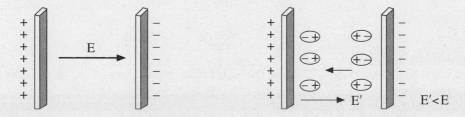

Fig. 1.30: Polarization of dielectric medium.

When the capacitor is charged, polarization of the molecules of the dielectric results. Polarization is a slight shift of negative charges towards the positive plate and of the positive charges towards the negative plate. This is due to the attraction or pull of unlike charges towards each other. The net charge on the plates in effect is thus reduced (Figure 1.30).

Due to this decrease in the charge on the plates the electric field is reduced from the value it would have had if there was no dielectric present.

Thus, if a dielectric is placed between two oppositely charged sheets of charge the electric field between the sheets decreases as the molecules of the dielectric get polarized.

1.18 Capacitors in Series and Parallel

As in the case of resistances, when capacitors are connected in series or in parallel the capacitance of the circuit is altered. This effective capacitance can be determined for capacitors connected in series and in parallel.

1.18.1 Capacitors in Series
When capacitors are connected in series the reciprocal of equivalent capacitance, Ceq, is equal to the sum of the reciprocal capacitances of the individual capacitors.

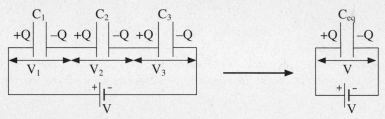

Fig. 1.31: Capacitors in series and equivalent capacitance.

In the circuit of Figure 1.31, three capacitors of capacitances C_1, C_2 and C_3 are connected in series to the source voltage V. Electrons from the negative terminal of the voltage source move to the right plate of C_3. A charge $-Q$ on this plate now induces a charge of $+Q$ on the left plate of C_3. This is because negative charge from the left plate of C_3 moved to the right plate of C_2. Again similar induction takes place and the plates of all three capacitors are charged. The source voltage V is equal to the sum of the voltages across the three capacitors. The charge on the other hand has the same magnitude Q on each capacitor plate. Here

$$V \quad = \quad V_1 \quad + \quad V_2 \quad + \quad V_3$$

$$Q/C_{eq} \quad = \quad Q/C_1 + \quad Q/C_2 + \quad Q/C_3 \qquad\qquad Q = CV$$
$$V = Q/C$$

$$\frac{1}{C_{eq}} \quad = \quad \frac{1}{C_1} \quad + \quad \frac{1}{C_2} \quad + \quad \frac{1}{C_3} \qquad\qquad\qquad 1.55$$

1.18.2 Capacitors in Parallel
When capacitors are connected in parallel the equivalent capacitance, Ceq, is the sum of the individual capacitances.

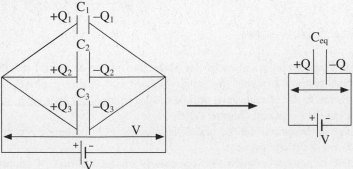

Fig. 1.32: Capacitors connected in parallel and the equivalent capacitance.

When capacitors are connected in parallel the voltage across each is V, the voltage of the battery. The charge Q divides between the capacitors according to their individual capacitances (Figure 1.32). Thus

$$Q \quad = \quad Q_1 \quad + \quad Q_2 \quad + \quad Q_3 \qquad\qquad Q = CV$$

$$C_{eq}V = \quad C_1V \quad + \quad C_2V \quad + \quad C_3V$$

$$\mathbf{C_{eq}} \quad = \quad \mathbf{C_1} \quad + \quad \mathbf{C_2} \quad + \quad \mathbf{C_3} \qquad\qquad\qquad 1.56$$

The total or equivalent capacitance of capacitors connected in parallel is equal to the sum of the individual capacitances.

1.19 Energy stored in a capacitor

A capacitor stores charge. In the circuit as shown in Figure 1.33, as soon as the key is switched on, a current flows in the circuit and the capacitor starts accumulating charge. The capacitor stops storing charge when the potential difference across its plates becomes equal to the voltage of the source (battery). Finally the same magnitude of charge Q is stored on each of the plates. This process takes place in a very short time.

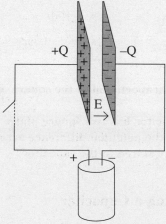

Fig. 1.33: The capacitor stores energy in the electric field E between the plates.

During the time the capacitor is charged, the voltage changes from 0 to V, so ΔV may be taken as the average p.d. over this time

$$\Delta V \quad = \quad V_{average} \quad = \quad \frac{(0 + V)}{2} \quad = \quad \frac{V}{2}$$

Energy stored is given by

$$\Delta U \quad = \quad Q\,\Delta V \qquad \text{(definition of p.d)}$$

$$\mathbf{\Delta U} \quad = \quad \frac{QV}{2} \qquad\qquad\qquad\qquad 1.57$$

As Q = CV

$$\text{Energy stored in capacitor} \quad = \quad \frac{CV^2}{2} \qquad\qquad 1.58$$

Note that this equation is analogous to the equation for Kinetic Energy

$$KE \quad = \quad \frac{mv^2}{2}$$

As the capacitor charges, a potential difference as well as an electric field is produced across the plates. Energy is stored in the electric field between the plates.

1.19.1 Energy is stored in the electric field of a capacitor

$$\text{Energy} \qquad = \qquad \tfrac{1}{2}\, CV^2$$

$$\qquad\qquad = \qquad \tfrac{1}{2}\, C\, E^2\, d^2 \qquad \text{as } V = Ed$$

$$\qquad\qquad = \qquad \frac{\tfrac{1}{2} A \varepsilon_o E^2 d^2}{d}$$

$$\text{Energy} \qquad = \qquad \tfrac{1}{2}\, A\, \varepsilon_o\, E^2 d$$

$$\text{Energy/unit volume} \quad = \quad \tfrac{1}{2}\, \varepsilon_o\, E^2 \quad \text{(since volume V = Ad)} \qquad 1.59$$

$$\textbf{Energy/unit volume} \quad \alpha \quad \mathbf{E^2}$$

Energy density is proportional to the square of the Electric intensity E.

Test yourself: A parallel plate capacitor has two square plates each of size 10 cm^2. The distance between the plates is 0.75 mm and the potential difference across the plates is 150 V. What is the charge on the plates and the energy stored in the capacitor?

1.20 Charging and Discharging a Capacitor

The behaviour of a capacitor when it is charged and discharged through a resistance will be discussed briefly in this section. How fast or how slow charging and discharging occurs depends on the magnitude of the resistance R and capacitance C.

1.20.1 Charging a Capacitor
An uncharged capacitor of capacitance C in series with a resistance R is connected to voltage source V as shown in the circuit diagram (Figure 1.34).

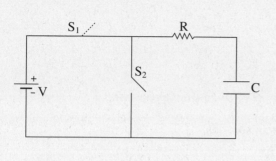
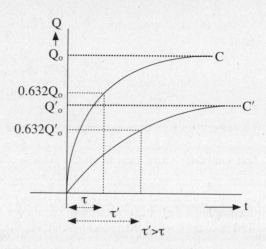

Fig. 1.34: Charging of a capacitor.

As the switch S_1 is closed at time t = 0, charges move from the voltage source and start collecting on the plates of the capacitor. At time t the capacitor is fully charged with charge Q_o and no charges will now flow in the circuit.

This buildup of charge on the capacitor takes place in a short time interval usually a few milliseconds. **The time in which the capacitor is charged to [1 − 1/e] or 0.632 of the maximum charge Q_o, it will finally attain what is termed as its time constant τ.** Value of e ≈ 2.718. It has been calculated that the time constant τ for a capacitor is equal to

$$\tau \quad = \quad RC \tag{1.60}$$

where R is the resistance and C the capacitance of the capacitor. If R is in ohms and C in farads, then τ is in seconds.

The equation for charging is $\mathbf{Q = Q_o\ [\ 1 - e^{-t/RC}\ ]}$ where Q is the charge on the capacitor at any time t.

Putting in the value of τ in equation $Q = Q_o\ [\ 1 - e^{-t/RC}\ ]$ the amount of charge accumulated in one time constant can be calculated:

$$
\begin{aligned}
Q \quad &= \quad Q_o\ [\ 1 - e^{-t/RC}\ ] \\
&= \quad Q_o\ [\ 1 - e^{-RC/RC}\ ] \\
&= \quad Q_o\ [\ 1 - e^{-1}\ ] \\
&= \quad Q_o\ [\ 1 - 0.368\ ] \\
&= \quad 0.632\ Q_o
\end{aligned}
\tag{1.61}
$$

From the graph, it can be seen that a capacitor with a smaller τ would charge faster than one with a larger τ.

The neuron behaves as a capacitor-resistance system. The myelin sheath is like a dielectric surrounded by conducting fluids (two parallel plates of a capacitor). The electric impulse encounters a resistance of about 13 MΩ as it flows through the axoplasm. The value of the capacitance in the axon is approximately 1.6 pF.

Thus, the time constant is $20\mu s$. The speed of the impulse can be calculated from the equation for speed

$$v \quad = \quad distance/time$$

$$v \quad = \quad 1mm/20\mu s \quad = \quad 50 \text{ m/s}$$

It has been found that nerve impulses travel at speeds of $60 - 90$ m/s in human axons.

Test yourself: Show that in the SI system of unit τ is in seconds.

Example: A $5\mu F$ capacitor, a $1\mu F$ capacitors and a $2M\Omega$ resistance are connected in series across a 120 V battery. Find the time constant and the charge on the capacitor at t = 3.34 s.

$$\text{Time constant} \quad = \quad RC_{eq}$$

$$1/C_{eq} \quad = \quad 1/C_2 + \quad 1/C_3$$

$$= \quad 1/(5\mu F) + \quad 1/(1\mu F)$$

$$C_{eq} \quad = \quad 0.833 \ \mu F$$

$$\text{Time constant} \quad \tau \quad = \quad RC_{eq}$$

$$= \quad 2M\Omega \times 0.833 \ \mu F$$

$$= \quad 1.67 \text{ s}$$

Charge on capacitor at t = 3.34 s:

$$Q \quad = \quad Q_o [1 - e^{-t/RC}]$$

$$= \quad Q_o [1 - e^{-3.34/1.67}]$$

$$= \quad Q_o [1 - 1/e^2] \qquad Q_o = C_{eq}V_o$$

$$= \quad C_{eq}V_o [1 - 0.135]$$

$$= \quad (0.833\mu \ F) (120V) (0.865)$$

$$= \quad 86.46 \ \mu C$$

1.20.2 Discharging a Capacitor

If now switch S_1 is opened and switch S_2 is closed, the capacitor will start discharging through R. It will completely discharge in a short time interval. **The time in which charge on the capacitor will drop to 1/e of its initial value Q_o is termed the time constant τ.** It is equal to

$$\tau \quad = \quad RC$$

where R is the resistance of the circuit and C the capacitance of the capacitor. The charge on the capacitor at time t is given by the equation

$$Q \quad = \quad Q_o\, e^{-t/RC} \qquad\qquad 1.62$$

where $Q_o = CV_o$ the initial charge.

The discharge curve is shown below in Figure 1.35. It discharges to 36.7% of the initial maximum value in one τ.

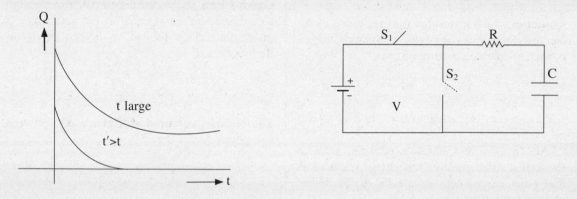

Fig. 1.35: Discharging of a capacitor.

For a capacitor with a smaller τ discharging would be faster and with a larger τ it would take more time. Again how fast the discharge occurs would depend on τ.

Summary

Electric charges have the following important properties:

1. Unlike charges attract one another and like charges repel one another.

2. Electric charge is always conserved.

3. Charge is quantized — it is in integral multiples of the charge on an electron proton.

4. The force between charged particles varies as the inverse of their separation.

Conductors are materials in which charges move freely. **Insulators** are materials that do not readily transport charge.

Coulomb's Law states that the magnitude of the electric force between two stationary charged particles separated by distance r is

$$F = \frac{k\, q_1\, q_2}{r^2}$$

k is a constant. It is equal to $k \approx 8.99 \times 10^9$ N.m^2/c^2

An electric field E exists at some point in space if a small positive test charge q placed at that point experiences an electric force F. The electric field is defined as

$$E = \frac{F}{q}$$

The direction of the electric field at a point in space is the direction of the electric force that would be exerted on a small positive charge placed at that point.

$$F = q\, E$$

The magnitude of the electric field due to a point charge q at distance r from the point charge is

$$E = \frac{kq}{r^2}\, \hat{r}$$

Electric field lines are useful for describing the electric field in any region of space. The electric field vector **E** is tangent to the electric field lines at every point.

The number of electric field lines per unit area through a surface perpendicular to the lines is proportional to the strength of the electric field in that region.

A **conductor** has the following properties:
1. The electric field is zero everywhere inside the conducting material.
2. On an irregularly shaped conductor, charge accumulates where the radius of curvature of the surface is smallest — at sharp points.

Gauss's Law states that the electric flux through any closed surface is equal to the net charge Q inside the surface divided by the permittivity of free space, ε_o:

$$\Phi = \frac{Q}{\varepsilon_o}$$

The **electric potential difference ΔV** between two points A and B is

$$\Delta V = V_B - V_A$$

$$= \frac{\Delta U}{q}$$

where ΔU is the change in electrical potential energy experienced by a charge q as it moves between points A and B. The units of potential difference are joules per coulomb, or **volts:**

$$\frac{1J}{1C} = 1V$$

The electric potential difference between two points A and B in a uniform electric field E is

$$V_B - V_A = -E\, d$$

where d is the distance between A and B and E is the strength of the electric field in that region.

The electric potential difference in a non-uniform electric field between two points A and B is

$$\Delta V = kq\left[\frac{1}{r_A} - \frac{1}{r_B}\right]$$

The absolute potential at a point r in space is

$$V = \frac{k\,q}{r}$$

Every point on the surface of a charged conductor in electrostatic equilibrium is at the same potential. The surface is called an equipotential surface.

The **electron volt** is defined as the energy that an electron (or proton) gains when accelerated through a potential difference of 1V. The conversion between electron volts and joules is

$$1\ eV = 1.60 \times 10^{-19}\ CV = 1.6 \times 10^{-19}\ J$$

A **capacitor** consists of two charged metal plates. The charge on each plate is equal in magnitude but opposite in sign. The capacitance C of any capacitor is the ratio of the magnitude of the charge Q on either plate to the potential difference V between them:

$$C = \frac{Q}{V}$$

Capacitance has the units coulombs per volt, or farad; 1 C/V = 1 F.

The capacitance of a parallel plate capacitor of area A separated by distance d is

$$C = \frac{\varepsilon_o A}{d}$$

where ε_o is a constant called the **permittivity of free space.** Its value is

$$\varepsilon_o = 8.85 \times 10^{-12}\ C^2/N.m^2$$

The **equivalent capacitance of a parallel combination** of capacitors is

$$C_{eq} = C_1 + C_2 + C_3 + \ldots\ldots$$

If two or more capacitors are connected in series, the **equivalent capacitance of the series combination is**

$$1/C_{eq} = 1/C_I + 1/C_2 + 1/C_3 \ldots\ldots$$

Three equivalent expressions for calculating the **energy stored** in a charged capacitor are

$$\text{Energy stored} = \frac{1}{2}\ Q\ (\Delta V)$$

$$= \frac{1}{2}\ C(\Delta V)^2$$

$$= \frac{Q^2}{2C}$$

When a non-conducting material, called a **dielectric**, is placed between the plates of a capacitor, the capacitance is multiplied by the factor ε_r, which is called the **dielectric constant or the relative permittivity** and is a property of the dielectric material. The capacitance of a parallel-plate capacitor filled with a dielectric is

$$C = \frac{\varepsilon_o \varepsilon_r A}{d}$$

Energy is stored in the electric field between the plates of a capacitor. In terms of the electric intensity it is

$$\text{Energy/unit volume} = \frac{1}{2}\ \varepsilon_o\ E^2$$

As a capacitor is charged by a battery through a resistor, the charge on the capacitor plates increases from zero to some maximum value. The **time constant $\tau = RC$** represents the time it takes the charge on the capacitor to increase from zero to 63% of its maximum value.

The equation for charging is $Q = Q_o\ [1 - e^{-t/RC}]$ where Q is the charge on the capacitor at any time t.

Questions

1. Are both bodies that attract each other electrically charged?
2. How can a neutral object acquire a negative charge?
3. Two positive charges P and Q are placed close together. They exert a repulsive force on each other. What would happen to the force if each of the charges is doubled?
4. Can the same effect be produced (Question 3) without changing the size of either charge? If so, how?
5. Why is the constant k included in the force equation?
6. Nearly all the mass of an atom is concentrated in its nucleus. Where is the charge located?
7. Why is the electric field intensity zero inside a charged conductor?
8. Why is charge generally transferred by electrons?
9. In what way did Millikan's experiment imply a minimum value of charge?
10. Which has more mass, an atom or an ion of the same element? Explain your answer.
11. Lines of force do not exist in reality. What is meant by them? Where do they originate? And what is their purpose?
12. Which unit is equivalent to coulomb^2 per newton-meter?
13. The dielectric constant of crown glass is 7. What does this mean?
14. Does charge have mass? Explain.
15. What is the difference between electrostatic force and gravitational force?
16. Show that the units for electric intensity volts/meter and newtons/coulomb are equivalent.
17. Does an electron move from a higher to a lower potential? Explain.
18. If you have three capacitors C_1, C_2 and C_3. How many different combinations of capacitance can be produced using all three?
19. What happens to the charge stored on a capacitor if the potential difference across the plates is doubled?
20. Many vehicles have windscreen wipers that are used intermittently when needed. How is this carried out by charging/discharging capacitors?

Problems

1. How many electrons can produce a charge of -500 μC?
2. How many electrons does a charge of 1nC represent?
3. The unit of electric intensity is
   a) J/C
   b) V/m
   b) C/m
   d) C/J
4. Two electric charges initially 8 cm apart are brought closer together until the force between them is greater by a factor of 16. How far apart are they now?
   a) 4 cm
   b) 2 cm
   c) 1.5 cm
   d) 1 cm
5. Two charges repel each other with a force of 10^{-6} N when they are 10 cm apart. What is the force between them when they are 2 cm apart?
   a) 2×10^{-6} N
   b) 2.5×10^{-5} N
   c) 1.5×10^{-6} N
   d) 2.5×10^{-6} N
6. What is the force between two small charges 40 μC and -10 μC placed 10 cm apart in air?
7. Two oppositely charged small spheres 1.2 cm apart in turpentine attract each other with the same force as when they are 1 cm apart in cotton seed oil. If the dielectric constant of turpentine is 2.2, what is the dielectric constant of cottonseed oil?
8. Where should a charge of 8 μC be placed between two charges 5 μC and 10 μC so that it experiences no force? The distance between the charges is 0.5m.
9. Two equal charges 5 cm apart repel with a force of 10^{-4} N. What are the charges?
10. What is the electric field intensity at a point 40 cm from a charge of 0.09 C in a vacuum?

11. Charges of 5 μC and 8 μC are placed 5m apart. What is the electric field at the midpoint between the charges?

12. What is the magnitude and direction of the force on q_2 due to q_1 and q_3?

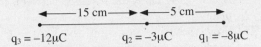

$q_3 = -12\mu C$ $q_2 = -3\mu C$ $q_1 = -8\mu C$

13. Find the force a charge q placed at the origin experiences due to the other two charges. (q = 0.5 μC)

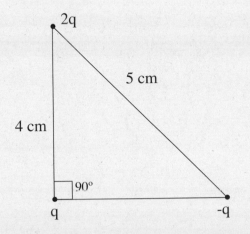

14. Point charges of magnitude q, 2q and 3q are placed at the corners of a square of side r. What is the magnitude and direction of the electric intensity at the fourth corner if q = 2 μC and r = 30 cm?

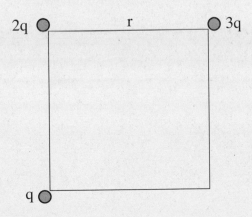

15. Two charges each 5 μC placed at A and B are separated by 50 cm. Find the electric intensity at P. Note PA = PB = 50 cm.

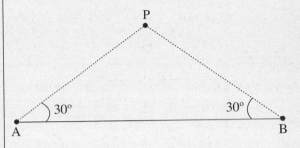

16. What electric field will keep a charge of – 15 μC suspended in air?

17. What must be the area of a tinfoil capacitor of 0.06 μF capacitance separated by paraffined paper (dielectric constant 2) and thickness of 0.025 mm?

18. If a parallel plate capacitor has a capacitance of 0.01 μF with its plates immersed in oil of dielectric constant 5 and a separation of 1 mm, what is the area of the plates?

19. Two capacitors are connected in series to yield a capacitance of 0.35 μF. If one has a capacitance of 0.5 μF, what is the capacitance of the other?

20. A parallel-plate capacitor has a capacitance of 1.2 nF and a charge of 0.8 C on each plate. a) What is the potential difference between the plates? b) The distance between the plates is doubled while the charge is kept the same, what will happen to the potential difference?

21. Find the equivalent capacitance.

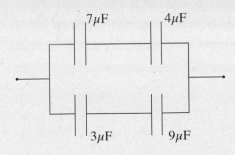

22. Find the equivalent capacitance across XY

    If $C_1 = 6\mu F$
        $C_2 = 12\mu F$
        $C_3 = 3\mu F$

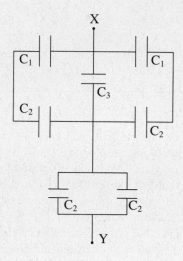

23. What is the velocity of a 50 eV proton?

24. An oil drop of mass 0.6×10^{-16} kg and charge 6.4×10^{-19} C is held stationary between the plates in Millikan's oil drop experiment. The voltage between the plates is 800 V. What is the distance between the plates?

25. Find the time constant for charging a conducting sphere having a capacitance of 3 μF and a resistance equal to 5000 Ω.

2 Current Electricity

OBJECTIVES

After studying this chapter, a student should be able:
- to understand currents and the requirement for producing a current.
- to understand that currents produce heating, chemical and magnetic effects.
- to understand resistance and its dependence on length, area, type of material and temperature.
- to understand the colour code for resistances.
- to understand about rheostats and potential dividers.
- to understand thermistors.
- to understand emf and internal resistance of voltage sources.
- to understand Kirchhoff's current and voltage laws and apply the laws to solve problems.
- to understand the principle of Wheatstone Bridge and how it can be used to find unknown resistances.
- to understand about potentiometer and its application.

2.1 Introduction

This chapter will focus on electric current. Electric current is flow of charge from one place to another. In other words electric current is moving charges. These moving charges transport energy. Electric current is one of the most convenient forms of energy transfer. Can you guess why?

When we need to transport gasoline from Karachi to Lahore or coal from Balochistan to Peshawar, it is carried in huge tankers/trucks. We have to go to the petrol station or godowns to buy it. On the other hand, when we use the hydroelectric power produced at Mangla Dam in homes in Lahore, we simply put on the electric switch and the bulb lights up or the fan starts rotating. Transfer of electric energy is carried by a pair of transmission wires which connect our homes to the source of electric power.

2.2 Electric Current

By definition **electric current I is the net amount of charge dQ moving through a cross-section in unit time.**

$$I \quad = \quad \frac{dQ}{dt} \qquad\qquad 2.1$$

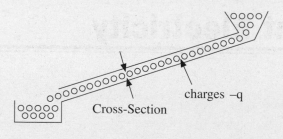

charges –q

Cross-Section

The unit for current is ampere. The current is one ampere when a net charge of one coulomb crosses a certain cross-section in one second.

$$1 \text{ ampere} = \frac{1 \text{ coulomb}}{1 \text{ second}}$$

$$= \frac{1C}{1s} = 1 \text{ A} \qquad\qquad 2.2$$

The symbol for ampere is A.

ΔQ

Fig. 2.1: Flow of charges through a cross-section.

Example: How many electrons flow in a light bulb in each second if the current passing through the bulb is 0.33 A?

$$I = 0.33 \text{ A}$$

We already know that a current of 1 A is equivalent to the charge carried by 6.25×10^{18} electrons crossing an area in 1 second.

Current of 1A $\equiv$ charge of 6.25×10^{18} electrons

Current of 0.33 A $\equiv$ $(6.25 \times 10^{18}) (0.33 \text{ A})$

$= 2.062 \times 10^{18}$ electrons flow in the light bulb each second

---

Current may be due to flow of electrons as in metallic conductors, flow of positive charges as in linear accelerators (Figure 2.2a) or flow of both types of charges as in electrolysis (Fig. 2.2b).

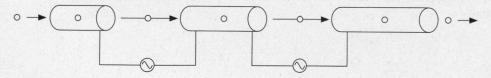

Fig. 2.2a: Flow of positive charges.

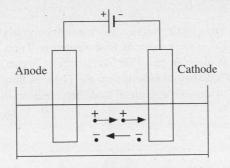

Fig. 2.2b: Electrolysis. Negative ions move to the left and positive ions to the right.

2.2.1 Conventional Current

By convention, the direction of current is taken as the direction of flow of positive charges. This means that when a circuit is connected to an energy source, for instance a battery, the current will flow from higher potential to lower potential. That is, it will flow from the positive terminal to the negative terminal of the battery (Figure 2.3a).

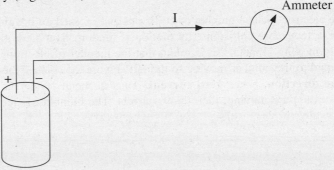

Fig. 2.3a: Current moves from positive terminal to negative terminal of a battery.

This can be done because the effect due to same amount of negative charges in one direction is equal to that of the same amount of positive charges in the opposite direction. In metallic conductors (wires) current is due to motion of free electrons but the direction of current is taken as the direction in which an equivalent positive charge would flow (Figure 2.3b). In this text, the conventional current practice shall be adopted.

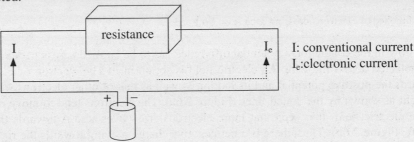

Fig. 2.3b: Conventional and electronic currents.

2.2.2 Current in metals

Metals have a crystalline structure. Their valence electrons are loosely bound to the atoms and can move easily around the metal. These are known as **free electrons**. They behave like gas molecules. In some texts, they are termed as an electron gas. The free electrons in metals move in straight lines until they collide with lattice ions. Then they change direction. There is thus a continuous random motion of electrons in metals. Figure 2.4 shows the path of one such electron in a metal.

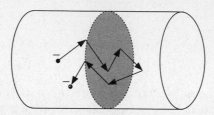

Fig. 2.4: Random motion of one electron (highly enlarged).

It can be seen that the electron made a number of collisions and crossed an imaginary cross-section. It re-crossed again after a few more collisions. Its motion is random. Hundreds of thousands of such electrons are randomly moving in metals. It is found that the number of electrons moving to the right through area A, are equal to the number moving to the left (Figure 2.5). Thus, there is **no net amount of charge in any one direction,** hence **zero current.** Thus in metals under normal circumstances, although charges (electrons) are moving, there is no current. The number of free electrons in copper wire is 10^{29}/m^3.

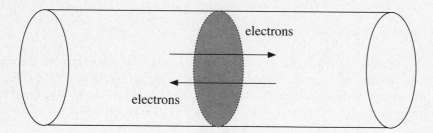

Fig. 2.5: Equal number of electrons crossing area A on both sides.

For current to flow in the conductor, a potential difference has to be applied across it. A voltage source (battery) is connected across the wire. An electric field is set up in it. The electron of Figure 2.4 will be pulled towards the positive potential and its motion as well as that of other electrons would be more towards the right as shown by the dotted lines (Figure 2.6a). The electrons tend to move in a direction opposite to the electric field. It is seen that more electrons cross over area A towards the right than towards the left (Figure 2.6b). Thus, there is a net negative charge moving towards the right. A current has been set up.

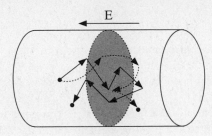

Fig. 2.6a: Electron moves against the electric field E.

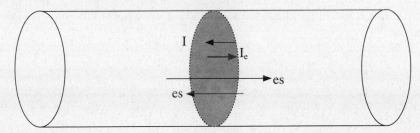

Fig. 2.6b: Net electron current is set towards the right (thick grey arrow). An equivalent conventional current is towards the left (black arrow).

As observed the actual current in the metallic conductor is from left to right but the conventional current (equivalent current due to positive charges) flows from right to left.

2.2.3 Drift velocity

In addition to the random motion, an overall small velocity **v** of the electrons towards the right is also noted. It is known as the drift velocity.

The equation for drift velocity can be easily derived. If n is the number of free electrons per unit volume and **v** the drift velocity, then n(Av) is the number that cross an area A per second. The velocity v is the distance travelled in one second.

The charge that crosses over per second is the current.

$$\text{Total charge} \quad Q \quad = \quad \text{(\# of charges) (charge on one electron)}$$
$$I \quad = \quad (n\,Av)\,(e)$$

where e is the charge on an electron. An electron carries a charge of 1.6×10^{-19} C.

$$\text{The drift velocity} \quad \mathbf{v} \quad = \quad \mathbf{I/nAe} \qquad\qquad 2.3$$

This velocity is about 10^{-3} m/s.

2.2.4 Conductors and Insulators

Conductors allow current to flow easily through them. As already discussed, metals have free electrons which can be made to flow by connecting the conductor to a voltage source. Insulators do not have free electrons. Their valence electrons are tightly bound to the atom and they are unable to conduct. Most metals are good conductors while materials like wood and rubber are poor conductors. Conductor and insulators will be discussed in Chapter 6 as well as semiconductors and superconductors.

2.3 Source of Current

To produce current there are two requirements:
  1) A path
  2) Potential difference

An analogy that can make this clear is the flow of water. You need to bring water from the WASA reservoir to your home. You would need a pipeline through which the water will flow from the reservoir to your house. You will also need to ensure that there is a difference in the water level. The level at the reservoir is higher than at your house otherwise the water will not flow. The pipeline is the wire or conductor (path) through which the charges flow and the water height difference is analogous to the potential difference.

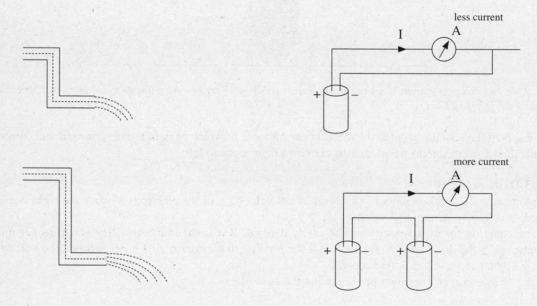

Fig. 2.7: Current increases as more batteries are connected.

If we want our water flow to increase we will have to increase the difference in height i.e. bring the water from a higher reservoir. Current behaviour is the similar. To increase the current the potential difference would have to be increased. This is depicted in Figure 2.7. What does height difference or potential difference imply? It means more potential energy. And it is that which drives the water as well as the current.

You are all familiar with cells and batteries that we use in torches, calculators, cameras and other portable devices. A torch cell lasts a few days. The energy stored in the cell is gradually used up in lighting the bulb. Therefore to maintain current a continuous source of energy is required.

How do we obtain this energy? We are aware that energy is conserved. It can be changed from one form to another. Electrical energy like other energy forms has to be converted from some other forms of energy.

What are the sources of electrical energy?

The sources of electrical energy are:

i)   Hydroelectric power stations convert the mechanical energy of falling water to electrical energy.

ii) Coal and gas powered plants convert chemical energy stored in coal and gas to electrical energy. Cells, batteries and generators also convert chemical energy to electric energy.

iii) Solar panels and solar cells convert solar energy to electrical energy.

2.4 Effects of Currents

Currents can be measured by their effects.

You must have noticed that a bulb gives out heat as well as light. Heater coils appear red hot and radiate heat. As currents flow through a wire, heat is produced. Some wires produce more heat than others. So current can be measured by its **heating effect.**

The amount of heat produced is given by

$$\text{Heat produced} \quad = \quad \text{(Power dissipated) (time)}$$

$$= \quad I^2Rt \qquad\qquad\qquad 2.4$$

where I is the current, R the resistance of the conductor and t the time.

The heat produced when current moves through a conductor is due to the collisions the electrons make with the lattice atoms. For current to flow a potential difference is applied across the conductor. Under the effect of the electric field so produced, the electrons accelerate, gaining in kinetic energy. They lose some of this energy as heat when they collide and are stopped or slowed down.

You have already studied the **chemical effects** of current in electrolysis. As current passes through an electrolyte, a chemical reaction occurs and the electrolyte breaks into positive and negative ions. Thus, currents produce chemical effects and can be measured by the amount of these chemical changes. As the current increases more conduction through the electrolyte takes place.

Two copper plates or rods are suspended in a beaker of copper sulphate solution. The rods are connected to a battery and an ammeter to measure current (Figure 2.8). The positively charged copper ions in the copper sulphate solution will be attracted to the negatively charged plate (cathode). The neutral copper atoms on the positive plate (anode) meantime dissolves into the solution. While dissolving they become positively charged leaving two electrons on the anode. These are the valence electrons.

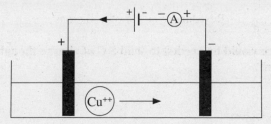

Fig. 2.8: The battery causes the positive copper rod to dissolve in the $CuSO_4$ solution and Cu^{++} ions are neutralized and deposited on negative plate.

When these positive ions reach the negative plate they pick up two electrons and deposit as copper atoms on the negative plate. The overall effect is the deposition of copper on the negative plate which comes from the positive plate. If a metallic spoon is used as the negative electrode it will be plated with copper.

Current flows in the circuit. In the electrolyte it flows from the positive electrode to the negative electrode. It is carried by the positively charged copper ions (Figure 2.8).

Magnetic effects are also associated with currents. This will be studied in more depth in Chapter 4. Magnetic fields are associated with currents. The magnetic field is directly proportional to the current. For different conductor configurations the geometry and material of the conductor may also affect the field produced. In general, the field is proportional to the current producing it:

$$\text{Magnetic field} \quad \alpha \quad \text{Current}$$

Magnetic effects of current can be useful in both detection and measurement of current. The current balance, an instrument to measure currents, uses this fact. Important electric devices like alternating and direct current generators and motors have been based on this effect.

2.5 Resistance

The concept of resistance is fundamental to flow of current. Resistance is opposition to the flow of current. It is defined as **the ratio of the potential difference applied across a conductor to the current produced.**

$$\text{Resistance} \quad = \quad \frac{\text{Potential difference}}{\text{Current}}$$

$$\mathbf{R} \quad = \quad \frac{\text{V}}{\text{I}} \qquad\qquad 2.5$$

The SI unit for resistance is volt/ampere or ohm (Ω). The conductor has a resistance of one ohm when a potential difference of one volt produces a current of one ampere through it.

$$\mathbf{1\ ohm} \quad = \quad \frac{1\ \text{volt}}{1\ \text{ampere}}$$

Physical resistance is due to the collisions between charge carriers (electrons in metallic conductors) and lattice atoms. **Resistance is due to collisions.** More collisions mean more resistance.

$$\text{Resistance} \quad \equiv \quad \text{\# of collisions}$$

Test yourself: What voltage would be needed to send 5 C of charge though a 10 Ω resistance in 2s? What would be the current?

2.5.1 Resistivity
Resistance depends on two factors:
- Dimensions of conductor
- Nature of conductor

Take two cylindrical conductors of equal areas but different lengths (Figure 2.9a top). More collisions will take place in the longer conductor of length ℓ' and it will offer more resistance. It is found that resistance is directly proportional to length:

$$\text{Resistance} \quad \alpha \quad \text{length}$$

$$\textbf{R} \quad \alpha \quad l \qquad\qquad 2.6$$

Now take two conductors of same length but different cross-sections (Figure 2.9b bottom). The one with the larger area A' has more spaces through which the electrons can move without colliding. It is found that the resistance is inversely proportional to the area of cross-section.

$$\text{Resistance} \quad \alpha \quad \frac{1}{\text{area}}$$

$$\textbf{R} \quad \alpha \quad \frac{1}{\textbf{A}} \qquad\qquad 2.7$$

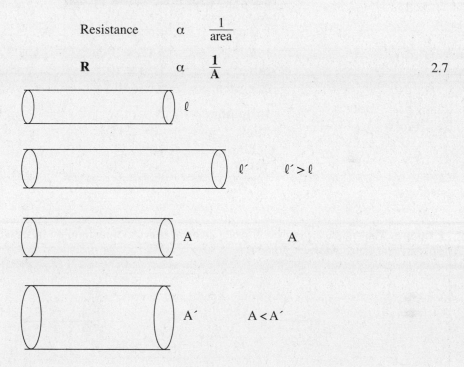

Fig. 2.9a Top and b Bottom: Resistance is directly proportional to length. Bottom: Resistance is inversely proportional to area of cross-section.

Combining equations 2.6 and 2.7

$$\textbf{R} \quad \alpha \quad \frac{\ell}{A}$$

$$= \quad \rho\frac{\ell}{A} \qquad\qquad 2.8$$

where ρ is known as the **resistivity.** In some texts, it is termed as **specific resistance.**
The resistivity is given by

$$\rho \quad = \quad \frac{R\,\ell}{A} \qquad\qquad 2.9$$

The resistivity ρ is the resistance offered by unit length and unit area of the conductor.

The unit of specific resistance is ohm - meter (Ω.m). By computing specific resistance comparison can be made between the resistances of different substances. Table 2.1 gives the specific resistance of some materials.

Thus the equation 2.8 for resistance R = ρL/A incorporates both dimensions and nature of the conducting material.

Table 2.1: Specific resistances of some materials

Material conductor	Resistivity Ω-m
Aluminium	2.82×10^{-8}
Copper	1.72×10^{-8}
Iron	9.17×10^{-8}
Silver	1.59×10^{-8}
Tungsten	5.6×10^{-8}
Insulators	
Mica	$10^{11} - 10^{15}$
Teflon	10^{16}
Wood	$10^{10} - 10^{13}$

Example: The conductor in the following diagram is made of aluminium (resistivity 2.75×10^{-8} Ω.m). Find its resistance between A and B.

Length of conductor l	=	0.6 m
Area of face A	=	(0.2 m) (0.2 m)
	=	0.04 m^2
Resistivity of Aluminium ρ	=	2.75×10^{-8} Ω.m
Resistance R	=	$\dfrac{\rho l}{A}$
	=	$\dfrac{(2.75 \times 10^{-8}\,\Omega.m)\,(0.6m)}{0.04\ m^2}$
	=	4.125×10^{-7} Ω

Test yourself: Compute the resistance of a copper wire of uniform density which has a length of 250 cm and a cross sectional area of 0.075 cm².

2.5.2 Temperature Coefficient of Resistance

Resistance is sensitive to temperature. With the rise of temperature resistance increases. The relation is linear over a wide range of temperatures for metallic conductors.

As temperature rises the lattice atoms vibrate faster and their amplitude also increases. Collision probability increases resulting in the increase of resistance. This is shown graphically in Figure 2.10.

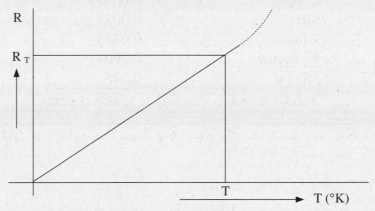

Fig. 2.10: Graph between temperature and resistance.

The temperature coefficient of resistance is defined as the fractional increase in resistance per unit rise in temperature. It is given by

$$\alpha = \frac{R_T - R_o}{R_o \, \Delta T} \qquad\qquad 2.10$$

Its unit in SI units is 1/K or K^{-1}.

When resistance increases with the rise of temperature α is positive. In some substances, resistance decreases with the rise of temperature α is negative. Conductors exhibit positive temperature coefficients of resistance. Some materials can have negative temperature coefficient of resistance. These include semiconductors like germanium whose resistance decreases when temperature rises.

Another equation for the temperature coefficient of resistivity is

$$\alpha = \frac{\rho_T - \rho_o}{\rho_o \, \Delta T} \qquad\qquad 2.11$$

as $R \propto \rho$ (equation 2.9). Here ρ_T and ρ_o are the resistivities at temperatures T and T_o respectively. ΔT is the temperature difference.

Equations 2.10 and 2.11 can be re-written as

$$\frac{R_T}{R_o} = \frac{\rho_T L/A}{\rho_o L/A} = \frac{\rho_T}{\rho_o}$$

$$\frac{R_T}{R_o} = \frac{\rho_T}{\rho_o} = 1 + \alpha \Delta T \qquad\qquad 2.12$$

Table 2.2 gives values of resistivities for some selected materials.

Table 2.2 Temperature Coefficients of Resistivity

Material	Temperature Coefficient of Resistivity $(C°)^{-1}$
Aluminium	0.0039
Carbon	−0.0005
Copper	0.00393
Silver	0.0038
Tungsten	0.0045
Nichrome*	0.0004
Silicon	−0.075

* A nickel-chromium alloy used in heating elements

Example: A nichrome wire has a resistance of 12 Ω when it is red-hot at 1200°C. What is its resistance at 25°C? ($\alpha_{nichrome} = 0.4 \times 10^{-3}$ °C^{-1})

$$R_T \quad = \quad 12 \ \Omega$$

$$T \quad = \quad 1200°C$$

$$T_o \quad = \quad 25°C$$

$$\alpha_{nichrome} \quad = \quad 0.4 \times 10\text{–}3°C^{-1}$$

$$R_o \quad = \quad ?$$

$$\Delta T \quad = \quad 1200°C - 25°C \quad = \quad 1175°C$$

$$\frac{R_T}{R_o} \quad = \quad 1 + \ \alpha \, \Delta T$$

$$R_o \quad = \quad \frac{R_T}{(1 + \alpha \Delta T)}$$

$$R_o \quad = \quad \frac{12}{\left[1 + \left(0.4 \times 10^{-3}°C^{-1}\right)\left(1175°C\right)\right]}$$

$$\quad = \quad 8.16 \ \Omega$$

2.6 Ohm's Law

Ohm's Law states that the current flowing in a conductor is directly proportional to the potential difference across it, provided temperature and other physical conditions remain constant.

$$\text{I} \qquad \alpha \qquad \text{V} \tag{2.13}$$

In metallic conductors, the graph between I and V is linear over a large range of temperature as can be seen in Figure 2.11. As the voltage becomes very large the graph deviates into a curve. A conductor that obeys Ohm's Law has a constant resistance over a wide range of potential difference. The straight line passes through the origin for ohmic conductors. These are conductors that obey Ohm's Law. Carbon conductors are non-ohmic. Non-ohmic conductors are those whose resistance changes with voltage or current.

The resistance for both ohmic and non-ohmic materials is given by V/I as discussed in section 2.4.

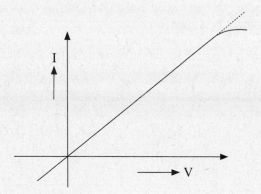

Fig. 2.11: I vs V for a metallic conductor.

It must be remembered that Ohm's Law is only valid for metallic conductors. The law does not hold for liquids, gases and semiconductors (Figure 2.12). In these current is not directly proportional to the voltage. When graphs are drawn between current and voltage for gases or liquids, the graph is not linear. This shows that Ohm's Law does not apply.

Ohm's Law unlike the conservation laws is not a fundamental law in physics.

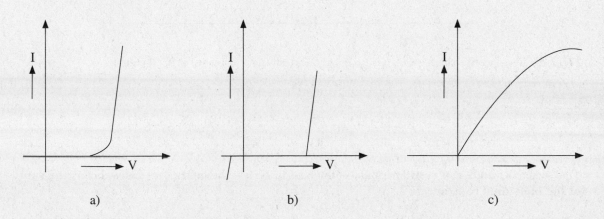

a)   b)   c)

Fig. 2.12: I vs V   a) gas in fluorescent tube light   b) acid   c) p-n junction

2.7 Series and Parallel Circuits

Electrical circuits have a number of resistances. The following sub-sections describe two basic types of resistance connections and how they can be computed.

2.7.1 Resistances in Series

In circuits there are usually more than one electric device connected to a voltage source. **Resistances are said to be connected in series when the same current flows through each resistance.**

Let three resistances be connected as shown in the Figure 2.13 to a voltage source. The voltage drop across each resistance depends on individual resistance and the sum of the voltages across each resistance is equal to the voltage V of the source.

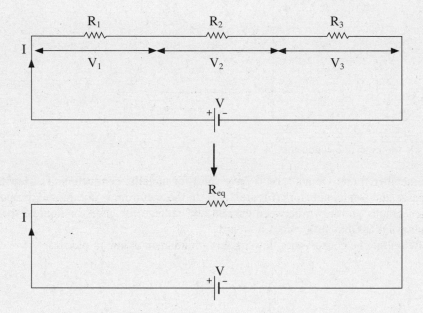

Fig. 2.13: Resistances connected in series (top) and their equivalent resistance is R_{eq} (bottom).

$$V = V_1 + V_2 + V_3$$
$$IR_{eq} = IR_1 + IR_2 + IR_3 \quad \text{as} \quad V = IR$$
$$\mathbf{R_{eq} = R_1 + R_2 + R_3} \qquad\qquad 2.14$$

Thus when **a number of resistances are connected in series, the equivalent resistance is the sum of the individual resistances.**

2.7.2 Resistances in Parallel

Parallel connections mean that the voltage across each electrical device (resistances) is the same.

The current divides according to the individual resistances. If resistance is more, less current flows in it. The current I from the battery will divide into currents I_1, I_2 and I_3 according to the value of the resistances R_1, R_2 and R_3 respectively (Figure 2.14).

$$I \quad = \quad I_1 \quad + \quad I_2 \quad + \quad I_3$$

$$V/R_{eq} = \quad V/R_1 + \quad V/R_2 + \quad V/R_3$$

$$1/R_{eq} = \quad 1/R_1 + \quad 1/R_2 + \quad 1/R_3 \qquad\qquad 2.15$$

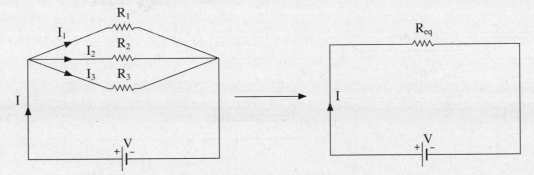

Fig. 2.14: Three resistances connected in parallel and their equivalent resistance R_{eq}.

When a number of resistances are connected in parallel, the reciprocal of their equivalent resistance will be equal to the sum of the reciprocals of the individual resistances.

If three resistances each equal to R are connected in parallel the equivalent resistance is R/3.

Example: If three resistances each equal to 60 ohms are connected in parallel as shown in the figure, what is the equivalent resistance?

$$R \quad = \quad 60\ \Omega$$

$$1/R_{eq} \quad = \quad 1/R \quad + \quad 1/R \quad + \quad 1/R \quad = \quad \frac{3}{60}$$

$$R_{eq} \quad = \quad \frac{60}{3}$$

$$= \quad 20\ \Omega$$

If n equal resistances are connected in parallel the equivalent resistance is R/n.

It can be worked out that if two resistances R_1 and R_2 are connected in parallel the equivalent resistance is

$$\frac{R_1 R_2}{R_1 + R_2} \qquad\qquad 2.16$$

Household connections are parallel connections. A 220 V supply is given to all household appliances as shown in Figure 2.15.

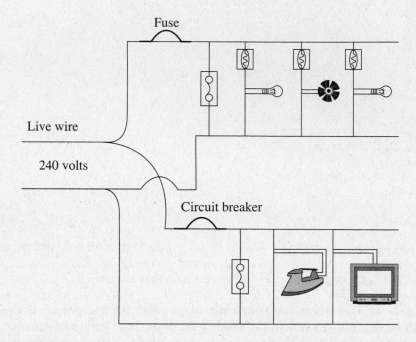

Fig. 2.15: Household circuit connections are parallel.

When circuits are connected in series and parallel equivalent resistances are determined separately for resistances observed in series and for those in parallel, these are replaced with their equivalent resistances. The circuit is now simplified. The circuit is then rechecked to find other series and parallel combinations till finally the circuit reduces to one single resistance. The following example would make this clear.

Example: Find the equivalent resistant of the given circuit?

In the circuit, R_1, R_2 and R_3 have the same voltage across them. They are parallel and their equivalent is

$$
\begin{aligned}
1/R_{eq1} &= 1/R_1 + 1/R_2 + 1/R_3 \\
1/R_{eq1} &= 1/5 + 1/10 + 1/20 \\
&= 7/20 \\
R_{eq1} &= 20/7 \\
R_{eq1} &= 2.85 \ \Omega
\end{aligned}
$$

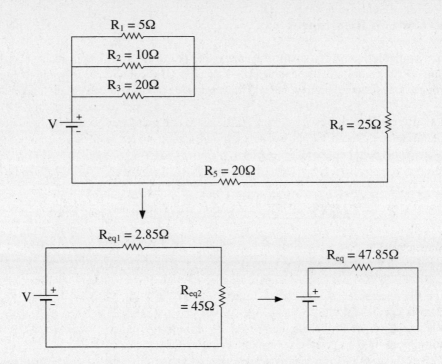

Resistances R_4 and R_5 are in series. Their equivalent resistance R_{eq2} is

$$R_{eq2} = R_4 + R_5$$
$$= 45 \ \Omega$$

The circuit reduces to two resistances in series. The final equivalent resistance is

$$R_{eq} = R_{eq1} + R_{eq2}$$
$$= 45 \ \Omega + 2.85 \ \Omega$$
$$= 47.85 \ \Omega$$

Test yourself: Find the equivalent resistance between A and B if R = 15Ω.
Note: Redraw to determine series and parallel connections.

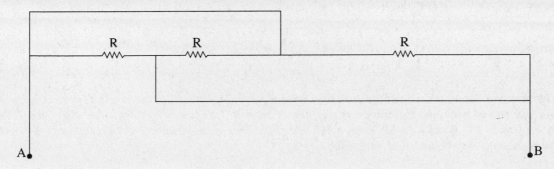

2.8 Colour Code in Resistances

Resistors are available from very small resistances like 0.001 Ω to very high resistances of the order of mega-ohms. Resistances are also characterized by their power measured in watts. If high current is passed through a resistance it may burn. Thus, wattage values are important and should not be exceeded.

There are different types of resistors. Wires of tungsten and manganin wound on insulators constitute one kind of resistance. Another kind is finely ground carbon mixed with insulator tightly packed in plastic capsules with metal edges (Figure 2.16).

Fig. 2.16: Some typical resistances.

A colour code (Table 2.3) has been designed to give the value of resistance. Colour bands or stripes are drawn on the resistance.

Bands 1 and 2 represent digits.

Band 3 is the power multiplier i.e. number of zeroes. For resistances below 10 Ω the third band is gold for multiplying by 0.1 and silver for multiplying by 0.01.

Band 4 gives the tolerance level. If it is silver the tolerance is ± 10%, if gold it is ± 5% and if pink it is 2%. If there is no band tolerance is ± 20%.

Table 2.3: Colour Code for Resistances

Colour	Value	Colour	Value
Black	0	Green	5
Brown	1	Blue	6
Red	2	Violet	7
Orange	3	Gray	8
Yellow	4	White	9

The bands are read from the side where they are nearest to the edge. The colour coding is standardized by Electronics Industrial Association (EIA).

For example, if resistance of the following resistor is to be estimated

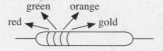

The first two bands are red and green. These are for the digits 2 and 5 respectively (Table 2.3). Band 3 is the power band and for colour orange the value is 3. The power by which the digits are to be multiplied is 10^3. Band 4 (gold) is for ± 5% tolerance. This means that 5% of the calculated value of the resistance may be added or subtracted.

red	green	orange	gold
2	5	10^3	± 5%

The resistance is

$$25 \times 10^3 \quad = \quad 25000 \ \Omega \pm [5\% \text{ of } 25000]$$
$$= \quad 25000 \ \Omega \pm [1250]$$
$$= \quad 26250 \ \Omega, 23750 \ \Omega$$

The resistance is between 26250 Ω and 23750 Ω.

Example: If a resistor has the following colour bands

Band 1: Red

Band 2: Green

Band 3: Gold

Band 4: Gold

What is the resistance of the resistor?

Band 1:   Red   digit 2

Band 2:   Green   digit 5

Band 3:   Gold   × 0.1

Band 4:   Gold   ±5%

Therefore resistance   =   [25 × 0.1] ± [5 % of 2.5]

=   2.5 ± 0.125

=   2.625 Ω, 2.375 Ω

Hence the resistance is between 2.375 Ω and 2.625 Ω

Test yourself: What resistance is equal to the following colour code?
Band 1: Brown, Band 2: Black, Band 3: Blue, Band 4: Silver.

2.9 Rheostat

In some circuits, resistances need to be varied. For instance, in dimmers of light or fans, resistances are gradually changed to alter current to give the desired light or speed in fans.

In the laboratory, a rheostat is such a device. The rheostat is a uniform coil. Resistance is proportional to length of coil. The coil can be used fully or partially in the circuit. As seen from the diagram, current enters from one of the fixed terminal (1) of the rheostat and leaves through the movable terminal (3).

As the knob is moved from left to right, the length of the coil through which the current has to flow increases and hence the resistance in the circuit is increased.

Figures 2.16 a) shows a schematic diagram of a rheostat and Figure 2.16 b) gives a circuit in which a rheostat is placed. The symbol for a rheostat is

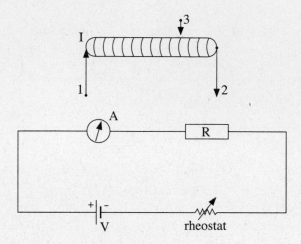

Fig. 2.16a and b (top): A schematic diagram of a rheostat.
(*bottom*): A circuit with a resistance, ammeter, battery and rheostat.

2.9.1 Potential Divider

The rheostat can also be used as a potential divider. In this way, a part of the potential difference V of the source can be obtained. The terminals (1) and (2) are connected across the voltage source V. The voltage V is across the total length of the coil. Then if a lesser value of voltage is required this can be obtained between terminal (1) and the movable terminal (3). This voltage will range from 0 to a maximum of V as the knob (3) moves from one end (1) to the other end (2) of the coil.

As the variable resistance between (1) and (3) increases, the voltage across the points also increases. Thus

$$V_{out} \quad \alpha \quad R_x \qquad \qquad 2.17a$$

where R is the resistance between (1) and (3)

and

$$V \quad \alpha \quad R \qquad \qquad 2.17b$$

Dividing 2.17a by 2.17b

$$\frac{V_{out}}{V} = \frac{R_x}{R}$$

$$\mathbf{V_{out}} = \frac{\mathbf{R_x}}{\mathbf{R}} \mathbf{V} \qquad \qquad 2.17c$$

The output voltage increases as R_x increases.

2.9.2 Thermistor

A thermistor is a temperature sensitive device. Its resistance decreases as temperature increases. A temperature dependent voltage can be obtained using thermistors as shown in Figure 2.17.

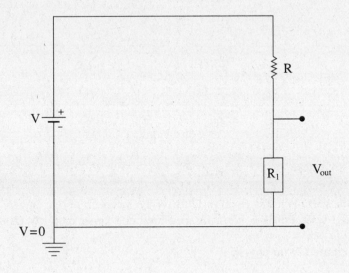

Fig. 2.17: A thermistor.

R is a fixed resistance while R_1 is the resistance of the thermistor. The output voltage V_{out} will be given by

$$V_{out} = \frac{R_1}{(R + R_1)} V \qquad\qquad 2.18$$

As temperature rises R_1 will decrease so will V_{out}. So the change in temperature can be detected by the variation in the voltage V_{out}.

Thermistors may also have positive temperature coefficients of resistance. In these resistance rises as temperature is increased. Negative temperature coefficient thermistors in particular can measure low temperatures accurately. They can be used as temperature sensors to detect minute changes in the temperature.

Thermistors are fabricated from ceramics in which oxides of nickel and cobalt are added.

2.10 Electrical Power

Power is rate of doing work.

As current moves through a resistance R, power loss in the form of heat loss takes place. Electric power is the amount of work done in maintaining a current I in the circuit. This amount of energy is to be continually supplied by the battery. To compute the power needed to maintain a current I let the voltage of the battery be V and resistance of the circuit R. Then power is given by

$$\text{Power} \quad = \quad \frac{\text{work done}}{\text{time interval}} \qquad\qquad 2.19$$

$$P \quad = \quad \frac{\text{energy}}{\text{time interval}}$$

$$= \quad \frac{\Delta U}{t} \qquad\qquad 2.20$$

$$= \quad \frac{qV}{t} \qquad \text{as } \Delta U = qV$$

$$= \quad I\,V \qquad\qquad \text{as } q/t = I \qquad\qquad 2.21$$

$$= \quad I^2\,R \qquad\quad \text{as } V = IR \qquad\qquad 2.22$$

$$= \quad V^2\,/R \qquad\qquad\qquad\qquad\qquad 2.23$$

Power supplied = power lost or dissipated

The unit for power is watt (W). It is equal to 1J/1s. When a current of one ampere flows across a p.d of 1volt, the power is 1 watt. Power is normally measured in a larger unit, kW (1000 W).

Energy can be determined from power.

From equation 2.20

$$\text{Energy} \quad = \quad \text{Power} \quad \times \quad \text{time}$$

$$\mathbf{\Delta U} \qquad = \quad \mathbf{IV\,t} \qquad\qquad\qquad 2.24a$$

$$= \quad \mathbf{I^2\,R\,t} \qquad\qquad\qquad 2.24b$$

We pay for the electrical energy by the amount of power we use per unit time into the length of time we use it for (equation 2.24a).

Another unit of energy besides joules and electron-volts is **watt-second.**

$$\mathbf{\Delta U} \quad = \quad \mathbf{P\,t} \qquad\qquad\qquad (2.18)$$

$$= \quad \mathbf{(1\ watt)(1\ second)}$$

$$= \quad \mathbf{1\ W.s}$$

The W.s is a small unit when we are dealing with consumption of electricity. A much larger unit, the **kilowatt-hour** is used for computing electrical energy consumed. **The kilowatt-hour is the electrical energy consumed at the rate of 1000 watts for one hour. It is equal to 3.6×10^6 joules.**

$$\mathbf{1\ kilowatt\text{-}hour} \qquad = \quad \mathbf{(1000\ W)\ (3600\ s)}$$

$$= \quad \mathbf{3.6 \times 10^6\ J} \qquad\qquad 2.25$$

Example: What is the cost of operating an electric heater that draws a current of 10 A on a 220 V circuit for 8 hours per day for a month? The electric rate is Rs 2.50 per unit.

$$I \quad = \quad 10 \text{ A}$$

$$V \quad = \quad 220 \text{ V}$$

$$\text{Time} \quad = \quad t \quad = \quad 8 \text{ hours/day} \times 30 \text{ days}$$

$$= \quad 240 \text{ hours}$$

$$\text{Rate} \quad = \quad \text{Rs 2.50 per unit (kWh)}$$

$$\text{Power} \quad = \quad V \, I$$

$$= \quad (220V)\,(10A)$$

$$= \quad 2200 \text{ W}$$

$$= \quad 2.2 \text{ kW}$$

$$\text{Electric energy consumed} \quad = \quad \text{Power} \times \text{time}$$

$$= \quad 2.2 \text{ kW} \times 240 \text{ hours}$$

$$= \quad 528 \text{ kWh}$$

$$1 \text{ kWh} \quad\quad \text{cost} \quad\quad \text{Rs} \quad 2.50$$

$$528 \text{ kWh} \quad\quad \text{cost} \quad\quad \text{Rs 2.50} \times 528$$

$$= \quad \text{Rs} \quad 1320$$

2.11 Electromotive Force

Electric current flows from a higher potential to a lower potential. To maintain this current, which would otherwise gradually become zero, a voltage source like a battery must be connected to the circuit. The battery or other source must continuously supply energy to the charge carriers to maintain the flow.

The energy that the battery supplies to a coulomb of charge is the electromotive force (emf). Its symbol is ε. The electromotive force is not a force but the energy that the battery provides to each unit (coulomb) of charge in driving it around the circuit.

In the SI system its unit is a volt. It is the energy in joules provided to each coulomb of charge to make it move round the circuit. Mathematically, the electromotive force is

$$\varepsilon \quad = \quad \frac{\text{Work done}}{\text{Unit charge}}$$

$$= \quad \frac{\Delta W}{q}$$

$$= \quad \frac{1J}{1C}$$

$$= \quad 1 \text{ volt (V)}$$

The battery/cell that you buy is capable of providing a potential difference of 1.5 V when no current is flowing through the circuit. Its emf is 1.5 V.

Besides batteries and cells, other sources of emf are generators, solar cells, hydroelectric power generators, etc. Emf sources convert different forms of energy like chemical energy, mechanical energy, solar energy into electrical energy.

2.11.1 Terminal Potential Difference

A battery of emf ε supplies a potential difference $V_{terminal}$ slightly less than ε to the circuit.

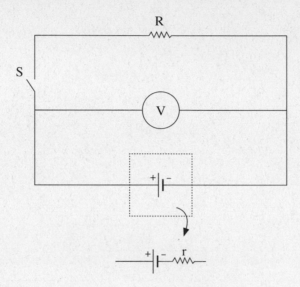

Fig. 2.18: Equivalent resistance r of the battery.

In Figure 2.18, when the switch S is closed a current I is drawn from the battery. As the current flows around the circuit, it also moves through the battery itself. The battery has a small internal resistance r. Some of the emf is utilized by the battery itself in moving the current through it. **The emf ε of the battery is the sum of the voltage drop due to current I flowing through the battery, $V_{battery}$, and the potential supplied to the battery terminals A and B, $V_{terminal}$, and thus to the circuit.**

$$\text{emf } \varepsilon \quad = \quad V_{battery} \quad + \quad V_{terminal}$$

$$\varepsilon \quad = \quad I\,r \quad + \quad I\,R \qquad\qquad 2.26$$

where r is the internal resistance of the battery and R the total resistance offered by the external circuit.

The current in the circuit is given by

$$I \quad = \quad \frac{\varepsilon}{R+r} \qquad\qquad 2.27$$

A voltmeter is attached to the battery to measure the potential difference. When no current is drawn, the voltmeter would show ε volts. But when the switch S is closed, the voltmeter reading will be less than ε and equal to $V_{terminal}$.

As discussed earlier, the battery provides the power for driving the current through the resistance. As the charges move through the resistance, energy is lost mainly as heat energy to the lattice atoms and surroundings. The power loss is $I^2 R$ (equation 2.22). This is the amount of power that must be supplied by the battery to maintain current I in the circuit of resistance R.

The power that the battery must provide is

$$\text{Power} = I^2 R$$

$$= \left(\frac{\varepsilon}{R+r}\right)^2 R \tag{2.27}$$

$$= \frac{\varepsilon^2 R}{R^2 + r^2 + 2Rr}$$

Re-writing

$$\textbf{Power} = \frac{\varepsilon^2 \textbf{R}}{(\textbf{R} - \textbf{r})^2 + \textbf{4Rr}} \tag{2.28a}$$

Thus the power that the battery can deliver depends on the emf ε of the battery, the internal resistance of the battery r and the resistance of the external circuit R. Power will be maximized when $\{(R - r)^2 + 4Rr\}$ is minimum i.e. when R = r:

Maximum power delivered by the battery

$$= \frac{\varepsilon^2 \textbf{R}}{\textbf{4Rr}} \tag{2.28b}$$

$$= \frac{\varepsilon^2}{\textbf{4r}} \tag{2.28c}$$

The power delivered by the battery is maximized when the circuit resistance and the battery's internal resistance are equal.

2.12 Kirchhoff's Rules

When circuits are complex and more than two voltage sources are incorporated as well as no two resistances are found in series or parallel, then it is not possible to simplify the circuits by series and parallel combinations. In such cases, Kirchhoff's current and voltage rules can be applied to solve the circuits.

Before discussing Kirchhoff's rules, certain nomenclature of circuits must be understood. Consider the circuit in Figure 2.19. Short circuit the batteries. The circuit has two loops and two junctions or nodes. The point where two or more wires or branches meet is termed as a junction or a node. The two nodes are a and d. The branches are connecting the two nodes. There are three branches abcd, ad and afed.

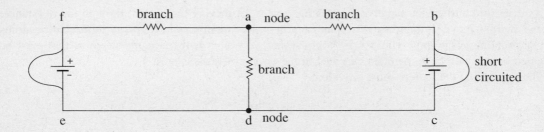

Fig. 2.19: A two loop circuit with two nodes and three branches.

2.12.1 Kirchhoff's First Rule
The current entering a junction must be equal to the current leaving the junction.

At a junction

$$\Sigma I = 0 \qquad\qquad 2.29$$

Kirchhoff's first rule is another way of stating the conservation of charge. Current or charge per unit time entering a junction is equal to that leaving a junction. No charge collects at the junction. What flows in flows out.

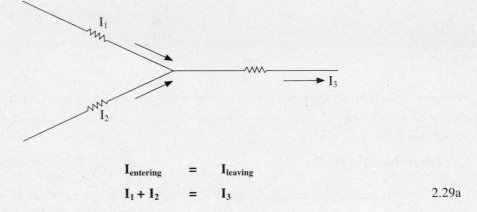

$$
\begin{aligned}
I_{entering} &= I_{leaving} \\
I_1 + I_2 &= I_3 \qquad\qquad 2.29a
\end{aligned}
$$

The solution of circuits by Kirchhoff's first rule is carried out in steps:

1) The number of junctions in the circuit are first determined after short circuiting the batteries.
2) The junction with the maximum number of nodes is assigned a voltage of zero. The other nodes are assigned voltage values V_1, V_2 ....
3) Kirchhoff's first law is now applied to each junction. It is assumed that the voltage at this junction is maximum and that all currents are moving away from the junction. Equations are so obtained for each junction/node.
4) The value of the voltage V_1, V_2 .... can be obtained by solving the equations.
5) If the number of junctions is n, the number of equations required to solve the problem is (n–1).

By applying the above rules we can solve the following circuit.

Example: Find the current in each resistor in Figure 2.20a.

The batteries are short circuited to determine the junctions. There are two junctions. Therefore number of equations to solve the problem:

$$n - 1 \ = \ (2 - 1) \ = \ 1$$

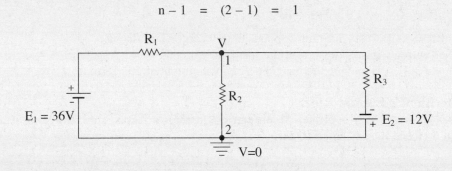

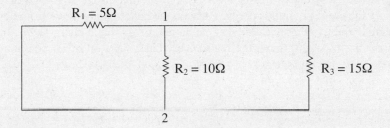

Fig. 2.20 a and b: Original circuit and circuit with batteries short-circuited.

Let voltage at junction **2** be zero and voltage at junction **1** be V. Replacing the batteries and applying Kirchhoff's Rule to junction **1**. Assuming all currents move away from junction **1**:

$$I_{entering} \ = \ I_{leaving}$$

$$0 \ = \ I_1 \ + \ I_2 \ + \ I_3$$

$$0 \ = \ \frac{V - E_1}{R_1} \ + \ \frac{V - 0}{R_2} \ + \ \frac{V - (-E_2)}{R_3}$$

$$0 \ = \ \frac{V - 36}{5} \ + \ \frac{V}{10} \ + \ \frac{V + 12}{15}$$

$$0 \ = \ 6(V - 36) + 3V + 2(V + 12)$$

$$0 \ = \ 6V - 216 + 3V + 2V + 24$$

$$192 \ = \ 11 \ V$$

$$V \ = \ 17.45 \ V$$

$$I_1 \quad = \quad \frac{17.45 - 36}{5} \quad = \quad -3.71 \text{ A}$$

$$I_2 \quad = \quad \frac{17.45}{10} \quad = \quad 1.745 \text{ A}$$

$$I_3 \quad = \quad \frac{17.45 + 12}{15} \quad = \quad 1.963 \text{ A}$$

The current $I_1 = -3.71$A. Its direction is opposite to the direction assumed.

The other currents move away from the junction. Currents 1.745 A and 1.963 A are entering the junction **1** and 3.71 A is leaving the junction. The total current at **1** is zero.

2.12.2 Kirchhoff's Second Rule

This is also known as the loop rule. According to Kirchhoff's second rule, **the sum of all voltage changes in a closed loop is equal to zero.**

$$\sum \Delta E \quad = \quad 0 \qquad\qquad 2.30$$

The voltage drop across the two resistances is from positive to negative. As the charges flow through the resistances they lose potential energy. Through the battery, they move from negative to positive and gain in potential energy (Figure 2.21). The potential energy that a charge gains in moving through the battery, it loses in moving through the resistances. Thus, the sum of all potential changes in a closed loop is zero.

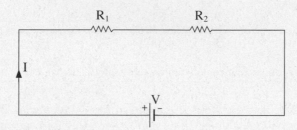

Fig. 2.21: Circuit for Kirchhoff's Loop Analysis.

The rules for solving circuits by the Loop Rule are:

1) Select the loops making sure all elements (resistances) in the circuit are included in the loops.
2) Assume currents in each loop.
3) Move around each loop in either clockwise or anti-clockwise direction. Each loop must be solved in the same direction — clockwise or anti-clockwise.
4) If the direction through a voltage source is from negative to positive, then the voltage change is taken as positive. If the voltage source is from negative to positive, then it is negative.
5) If the direction through a resistance is the same as the direction of the current, then the voltage change is taken as negative. If it is opposite to the current direction it is taken as positive. All voltage changes through one loop must have the same polarity.
6) Formulate equations for each loop applying Kirchhoff's second rule and solve. The loop being considered is assumed to have the highest current as compared to the other loops.

Example: Determine the currents in R_1, R_2 and R_3 using Kirchhoff's second rule.

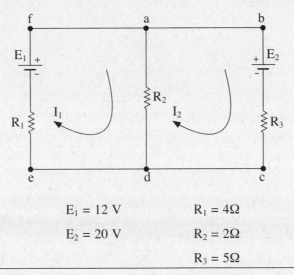

$$E_1 = 12 \text{ V} \qquad R_1 = 4\Omega$$
$$E_2 = 20 \text{ V} \qquad R_2 = 2\Omega$$
$$R_3 = 5\Omega$$

The loops are selected ensuring that each resistance is in either one or the other loop. Current in loop **adefa** is I_1 in the clockwise direction. Current in loop abcda is I_2 and in the clockwise direction. When applying Kirchhoff's rules, the current in the loop under consideration is greater than current in the other loops, starting from **a** and moving in the clockwise direction.

$$-(I_1 - I_2)\, R_2 \quad - \quad I_1 R_1 \; + \quad E_1 \quad = \quad 0 \qquad\qquad 2.30a$$

As it can be seen from the above figure, the current in R_2 is $(I_1 - I_2)$. Now applying Kirchhoff's second rule to loop **abcda** starting from **a** again in the clockwise direction.

$$-E_2 \quad - \qquad I_2 R_3 \; - \qquad (I_2 - I_1)\, R_2 = \quad 0 \qquad\qquad 2.30b$$

Putting values in equation 2.30a and simplifying

$$-(I_1 - I_2)\, 2 - \qquad I_1\,(4) \qquad + \qquad 12 \quad = \quad 0$$
$$-6I_1 \qquad + \qquad 2I_2 \quad = \quad -12$$
$$-3I_1 \qquad + \qquad I_2 \quad = \quad -6 \qquad\qquad 2.30c$$

and equation 3.30b becomes

$$-20 \quad - \qquad 5I_2 \quad - \qquad (I_2 - I_1)\, 2 \quad = \quad 0$$
$$-20 \quad - \qquad 7I_2 \quad + \qquad 2I_1 \qquad = \quad 0$$
$$2I_1 \quad - \qquad 7I_2 \qquad\qquad = \quad 20 \qquad\qquad 2.30d$$

Solving 2.30c and 2.30d

$$I_1 \quad = \quad 1.158 \text{ A clockwise}$$

$$I_2 \quad = \quad -2.526 \text{ A anti-clockwise}$$

The current through R_2 and R_3 is 3.685 A and 2.526 A respectively in the anti-clockwise direction whilst the currents in R_1 is 1.158 A in the clockwise direction.

2.13 Wheatstone Bridge

The Wheatstone Bridge circuit was designed to measure unknown resistances. The Wheatstone Bridge is a circuit with four resistances connected as shown in Figure 2.22. A galvanometer G of resistance R_g is connected between B and D and the voltage source E is between A and C.

The Wheatstone Bridge relationship between the four resistances has to be determined when no current is registered by the galvanometer G.

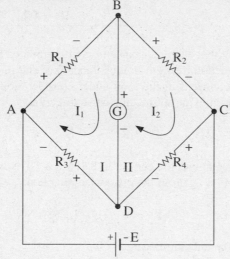

Fig. 2.22: The Wheatstone Bridge.

To determine this relation between the four resistances, Kirchhoff's second rule is applied. There are 3 loops in the circuit but the loops I and II need to be solved to determine the relationship between R_1, R_2, R_3 and R_4 when no current flows through the galvanometer G.

Applying Kirchhoff's second rule to loop I

$$-I_1 R_1 - \quad (I_1 - I_2) R_g - \quad I_1 R_3 \quad = \quad 0 \qquad\qquad 2.31a$$

And to loop II

$$-I_2 R_2 - \quad I_2 R_4 + \quad (I_2 - I_1) R_g = \quad 0 \qquad\qquad 2.31b$$

Applying the condition for the working of the Wheatstone Bridge i.e. no current should flow in the galvanometer G. This is achieved when the potential at B and D is equal or

$$(I_1 - I_2) R_g \quad = \quad 0 \qquad\qquad 2.31c$$

which means $$I_1 - I_2 \quad = \quad 0$$

Let $$I_1 \quad = \quad I_2 \quad = \quad I$$

The equations 2.31a and 2.31b become

$$-I R_1 \quad = \quad IR_3 \qquad\qquad 2.31d$$

and $$-I R_2 \quad = \quad IR_4 \qquad\qquad 2.31e$$

Dividing 2.31d by 2.31e

$$\frac{R_1}{R_2} \quad = \quad \frac{R_3}{R_4} \qquad\qquad 2.32$$

Equation 2.32 is known as **the principle of the Wheatstone Bridge. It states that when no current passes through the bridge and galvanometer, the ratio of the resistances R_1 and R_2 is equal to that of R_3 and R_4.**

This principle is important in determining the unknown resistances.

In Figure 2.23, X is the unknown resistance. Resistances P and Q are fixed. The variable resistance S is changed till no current passes through the galvanometer G. The Wheatstone Bridge condition is fulfilled.

Then the unknown resistance is $\quad \mathbf{X} \quad = \quad \left(\dfrac{P}{Q}\right) S$ $\qquad\qquad 2.33$

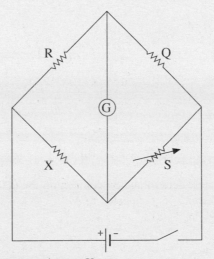

Fig. 2.23: Circuit to determine unknown resistance X.

2.14 Potentiometer

The potentiometer circuit is designed to measure the potential difference with accuracy.

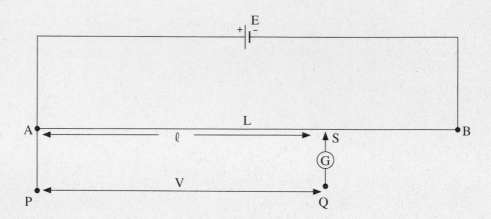

Fig. 2.24: A Potentiometer.

A straight wire AB of uniform cross-section, resistance R and length L is connected to a battery E as shown in Figure 2.24. Being a wire of uniform cross-section, its resistance per unit length is the same throughout the length of the wire. The voltage V across P and Q is thus proportional to the length where the sliding contact S is:

$$V \quad \alpha \quad \ell$$
$$V \quad = \quad k\ell \qquad\qquad\qquad 2.34a$$

where k is the constant of proportionality.

Similarly

$$E \quad \alpha \quad L$$
$$E \quad = \quad kL \qquad\qquad\qquad 2.34b$$

Comparing equations 2.34a and 2.34b

$$V \quad = \quad E\left[\frac{\ell}{L}\right] \qquad\qquad\qquad 2.35$$

The emf of a cell can be accurately determined by connecting the cell between P and Q (Figure 2.25) and ensuring that the positive terminal is connected to A.

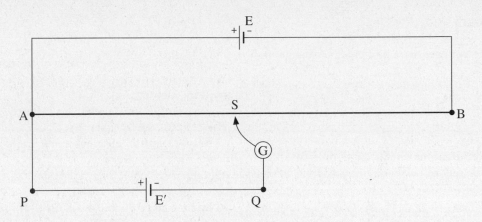

Fig. 2.25: Circuit to determine emf E′ for a cell.

When connected, the galvanometer G will show a deflection. The sliding contact S is tapped along the wire AB until the galvanometer shows zero deflection. It does not draw any current. Then the emf E of the cell is equal to the voltage across AS. Thus the emf E′ can be accurately determined as no current has been drawn from the cell.

Using equation 2.35, an unknown emf E′ can be determined as

$$E \quad \alpha \quad \ell \qquad\qquad\qquad 2.36$$

and

$$E' \quad \alpha \quad \ell' \qquad\qquad\qquad 2.37$$

Dividing 2.36 by 2.37

$$\frac{E}{E'} = \frac{\ell}{\ell'}$$

$$E' = E\left[\frac{\ell'}{\ell}\right] \qquad\qquad\qquad 2.39$$

Care should be taken that the emf E′ < E as the potentiometer cannot measure voltages greater than E.

Voltmeters are instruments to measure potential difference. They have very high resistances so that they hardly draw any current from the circuit. Still they do draw a very small current that alters the circuit current and the potential difference that is being measured.

Summary

The **electric current** I in a conductor is defined as

$$I = \frac{dQ}{dt}$$

where ΔQ is the charge that passes through a cross-section of the conductor in time Δt.

The SI unit of current is the ampere (A):

$$1A = 1 \text{ C/s}$$

By convention, the direction of current is the direction of flow of positive charge. The current in a conductor is related to the motion of the charge carriers by

$$\mathbf{I = nAev}$$

where n is the number of mobile charge carriers per unit volume, q is the charge on each carrier, v is the drift speed of the charge, and A is the cross-sectional area of the conductor.

The resistance R of a conductor is defined as the ratio of the potential difference across the conductor to the current:

$$R = \frac{\Delta V}{I}$$

The SI units of resistance are volts per ampere, or **ohms** (Ω):

$$1\Omega = 1 \text{ V/A}$$

Ohm's law describes metallic conductors for which the applied voltage is directly proportional to the current it causes. The proportionality constant is the resistance:

$$V = IR$$

If a conductor has length ℓ and cross-sectional area A, its **resistance** is

$$R = \frac{\rho \ell}{A}$$

where ρ is an intrinsic property of the conductor called the **resistivity**, The SI unit of resistivity is the **ohm-meter** (Ω.m).

Resistances are labeled by coloured bands. The colour code is given by:

Bands 1 and 2 represent digits. **Band 3** is the power multiplier. For resistances below 10 Ω the third band is gold for multiplying by 0.1 and silver for multiplying by 0.01.

Band 4 gives the tolerance level. If it is silver the tolerance is $\pm$ 10%, if gold it is $\pm$ 5% and if pink it is $\pm$ 2%. If there is no band, tolerance is $\pm$ 20%.

The **resistivity** of a conductor varies directly with temperature over a limited temperature range according to the expression

$$\rho = \rho_0 [1 + \alpha (T - T_0)]$$

where α is the **temperature coefficient of resistivity** and ρ_0 is the resistivity at some reference temperature T_0 and $T - T_0$ is the change in temperature.

The resistance of a conductor varies with temperature according to the expression

$$R = R_0 [1 + \alpha (T - T_0)]$$

If a potential difference V is maintained across an electrical device, the **power**, or rate at which energy is supplied to the device is

$$P = IV$$

Because the potential difference across a resistor is V = IR, the **power delivered to a resistor** can be expressed as

$$P = I^2 R = \frac{V^2}{R}$$

A **kilowatt-hour** is the amount of energy converted or consumed in 1 hour by a device supplied with power at the rate of a kW. This is equivalent to

$$1 \text{kWh} \ = \ 3.6 \times 10^6 \text{ J}$$

Maximum power delivered by the battery

$$P_{max} \ = \ \frac{E^2}{4r}$$

A source of emf is any device that transforms non-electrical energy into electrical energy.

The **emf and terminal potential difference $V_{terminal}$** are related by

$$\varepsilon \ = \ I (r + R)$$
$$= \ I r + V_{terminal}$$

The **equivalent resistance** of a set of resistors connected in series is

$$R_{eq} \ = \ R_1 + R_2 + R_3 + \dots\dots$$

The **equivalent resistance** of a set of resistors connected in **parallel** is

$$1/R_{eq} \ = \ 1/R_1 + 1/R_2 + 1/R_3 + \dots\dots$$

The complex circuits are conveniently analyzed by using **Kirchhoff's rules:**
1. The sum of the currents entering any junction must equal the sum of the currents leaving that junction.
2. The sum of the potential differences across all the elements around any closed circuit loop must be zero.

The first rule is a statement of **conservation of charge.** The second rule is a statement of **conservation of energy.**

When the Wheatsone Bridge is balanced, the 4 resistances are related by

$$\frac{R_1}{R_2} \ = \ \frac{R_3}{R_2}$$

The potentiometer circuit measures the potential difference with accuracy. The mathematical expression for a potentiometer is

$$\frac{E}{E'} \ = \ \frac{\ell}{\ell'}$$

Questions

1. Do free electrons move the full length of the wire when under the influence of a potential difference?
2. Are all liquids good conductors of electric current?
3. Upon what factors does resistance of a conductor depend?
4. What is the difference between emf and electric potential? What are the units of each?
5. Distinguish between terminal voltages of a battery a) when delivering current b) when there is no current.
6. What is meant by a closed electrical circuit? Will charge flow in an open circuit? Explain.
7. If current is doubled and voltage halved, will the heat produced in unit time remain constant? Defend your answer.
8. If a current in a conductor is doubled what would happen to the drift velocity?
9. What conditions are necessary for currents?
10. When the voltage across a conductor is doubled the current is found to have become four times. What can you say about this conductor?
11. When we change the speed of a fan what do we change?
12. Two wires of the same material and same length have different resistances. One wire has a resistance equal to three times that of the other. Can you suggest why this is so?
13. What path does the current take through resistors connected in series?

14. If N resistances of R Ω each are connected in parallel, what is their combined resistance?

15. Why does the bulb filament get hot while the wire that goes to the bulb does not get hot although the same current goes through both?

16. Why does resistance rise when the temperature increases?

17. How can you connect resistances so that the equivalent is larger than the individual resistances?

18. You are given two sets of ornamental lights to be placed on the walls of your house for your uncle's wedding. In one set, when one bulb is removed the others remain lighted, in the other, when one bulb is removed the remaining do not light up as well. What is the difference between the connections of the two sets?

19. Would a fuse work successfully if it was connected in parallel to the device it was supposed to protect?

20. Can a voltmeter measure the emf of a cell?

21. When a Wheatstone Bridge is connected why is there no current through the galvanometer?

22. Why is the potentiometer used to measure voltage?

23. Distinguish between a rheostat and potentiometer.

24. Upon what does conduction of current through a liquid depend?

25. What is the unit for electrical energy?

Problems

1. Three resistances 1 ohm, 5 ohms and 10 ohms are connected in parallel. The resulting resistance is
   a) Greater than 16 ohms
   b) Equal to 16 ohms
   c) Equal to 1.3 ohms
   d) Less than 1.3 ohms

2. A piece of copper has maximum resistance when its area and length are
   a) A and L
   b) A/2 and 2L
   c) 2A and L/2
   d) 3A and 2L

3. Fill in
   i) 0.75 A   = _____ mA
   ii) 8 $\mu\Omega$ = _____ Ω

4. What amount of charge will flow through a conductor if a current of 4 A is maintained for 10 hours?

5. A current of 3.5 A flows through a wire. How much charge passes through the wire every minute?

6. In a wire maintained at 120 V, a current of 2 A flows and in another maintained at 240 V the same current flows. What are the rates at which charges flow though each wire?

7. A platinum wire 80 cm long is to have a resistance of 0.1Ω. What should its diameter be? (The resistivity of platinum is 1.1×10^{-7} Ω–m).

8. A rectangular block of copper has sides of length 10 cm, 20 cm, 40 cm. If connected to a 6 V source across opposite faces of the rectangular block, what is
   a. the maximum current
   b. the minimum current that can be carried?

9. A potential difference of 12 V is found to produce a current of 0.4 A in 3.2 m length of wire with a uniform radius of 4 cm. What is
   a. the resistance of wire
   b. the resistivity of the wire

10. A wire has a resistance of 400 ohms. What will be the resistance of wire of the same material if it is 4 times as long and its area of cross-section is 2 times as large?

11. A wire has a resistance of 21Ω. It is melted down and from the same volume of metal a new wire is made that is 3 times longer than the original wire. What is the resistance of the new wire?

12. There are four coils of resistances 30 Ω, 50 Ω, 18 Ω, and 36 Ω respectively. The first two are connected in parallel and the last two are also connected in parallel. The two pairs are then connected in series across a 220 V battery. Find the current in each coil.

13. What resistance must be connected in parallel to a 60 Ω coil to make the combined resistance 40 Ω?

14. Draw a circuit containing resistances of 100 Ω, 25 Ω and 70 Ω which would give an equivalent resistance of 90 Ω.

15. The resistance between terminals A and B is 65 Ω. If the resistors labeled R have the same value, find R.

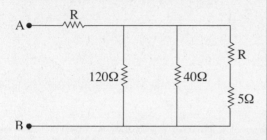

16. Find the equivalent resistance and the currents I and I_1.

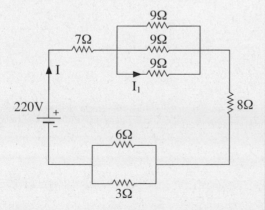

17. Find I_1, I_2 and I_3 using Kirchhoff's junction rule.

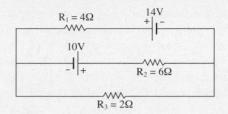

18. Using Kirchhoff's Law find the currents in R_1, R_2 and R_3.

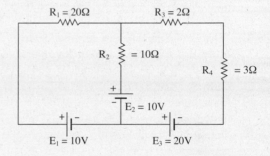

19. A bird is sitting on a high voltage power line. The copper wire on which it is sitting is 2.2 cm in diameter and carries a current of 50 A. If the bird's feet are 4 cm apart, calculate the potential difference across its feet.

20. A tungsten wire has a resistance of 100 Ω at 220°C. If the resistance increases 0.005 Ω/Ω/°C as the temperature rises, what is the resistance at room temperature (20°C)?

21. A cloud is at a potential of 8×10^6 V relative to the ground. A charge of 40 C is transferred in a lightning stroke between cloud and the ground. Find the energy dissipated.

22. A 2.4 kW generator delivers 10A. At what potential difference does the generator operate?

23. How many 100 W bulbs can you use in a 240 V circuit without tripping a 15 A circuit breaker? The bulbs are connected in parallel which means that the potential difference across each light bulb is 240 V.

24. What is the ratio of the resistances of two bulbs A and B marked 60 W, 110 V and 60 W, 220 V respectively?

25. A battery has an emf of 6.6 V and an internal resistance of 0.5 Ω. It is connected to a coil having a resistance of 0.5 Ω. What is the reading of the voltmeter connected to the battery?

26. A battery is delivering 10 A to a resistance of 0.6 Ω. There is no other resistance in the external circuit. If the internal resistance of the battery is 0.02 Ω, what is its emf?

27. You are watching a cricket match for 8 hours on a 180 W television set? How much would it cost to view the match at Rs 3.50/kWh?

28. What would it cost to toast two slices of bread if it requires 8 mins with a current of 6 A through 20 Ω resistance? (One kWh costs Rs 4.20).

29. How much current is drawn by a 1.5 hp motor operating at 220 V?

30. Determine the following resistances using colour code

    i) Band 1   orange
       Band 2   blue
       Band 3   red
       Band 4   silver

    ii) Band 1   black
        Band 2   brown
        Band 3   green
        Band 4   gold

31. Give the colour code for a 0.65 Ω resistance with 5% tolerance.

32. Three resistances 20 Ω, 30 Ω and 50 Ω are connected in series across a 240 V battery. What is the voltage across the 30 Ω resistance?

33. A person experiences a mild shock if the current across the thumb and index finger is more than 80 μA. Compare the maximum allowable voltage without shock across the thumb and index finger with a dry-skin resistance of 4×10^5 Ω and a wet-skin resistance of 2000 Ω.

3 Electromagnetism

OBJECTIVES

After studying this chapter, a student should be able:
- to understand that magnetic field is associated with moving charges.
- to understand a moving charge or current in a magnetic field experiences a force.
- to understand that magnetic flux is similar to electric flux.
- to understand Ampere's circuital law and its application in computing the magnetic induction.
- to understand that a charged particle in an electric and magnetic field experiences a force.
- to understand that charged particles move in circular paths when projected perpendicular to the magnetic field.
- to understand the importance of e/m and its measurement.
- to understand how the cathode ray oscilloscope is useful in visualizing and measuring voltages and currents.
- to understand the torque on a current-carrying conductor.
- to be able to understand about galvanometer and how it can be converted into an ammeter, a voltmeter and an ohmmeter as well as multi-range meters.

3.1 Introduction

Man has been familiar with magnets since ancient times. The lodestone (magnetite) and its usefulness have been known to man for more than a thousand years.

Electricity and magnetism were thought to be different phenomena. It was only in the 18th century that scientists realized that the two were related. Electromagnetism as the name suggests is the magnetic effect due to electric charge. Associated with the moving charge is a magnetic field whilst stationary charges produce electric forces. Thus, when charges move across magnetic field the two fields interact and their motion is affected. **Magnetic field and forces are associated with moving charges.**

When charges are stationary, the branch of electricity dealing with them is termed as electrostatics and when charges are in motion, the forces and fields are termed as magnetic forces and fields. Magnetic fields are very important in the production of electricity, in motors and in different applications in industry and medicine.

In this chapter, the forces and effects of magnetic fields, and their applications will be discussed.

3.2 The Magnetic Field Associated with a Current in a Long Straight Wire

Take a long, straight wire with a current I flowing through it. It is found that a magnetic field surrounds this wire in concentric circles. As you move away from the conductor, the field decreases inversely with distance.

$$B \quad \alpha \quad \frac{I}{r} \qquad \qquad 3.1$$

When current is doubled the field is also twice as much. It varies directly with the current I

$$B \quad \alpha \quad I \qquad \qquad 3.2$$

Combining equations 3.1and 3.2 the field equation becomes

$$B \quad \alpha \quad \frac{I}{r}$$

$$= \quad \frac{\mu_o I}{2\pi r} \qquad \qquad 3.3$$

μ_o is the constant of proportionality and is called the permeability of free space and is equal to $4\pi \times 10^{-7}$ T.m/A. The magnetic field depends on the medium.

This expression is similar to that of the electric intensity $(E = kq/r^2)$ except that the electric field decreases with the square of the distance from the source (r^2).

The direction of the magnetic induction B can be found by the right hand rule. Grasp the right hand around the straight conductor so that the thumb points in the direction of the current. The fingers would curl around along the magnetic field direction (Figure 3.1).

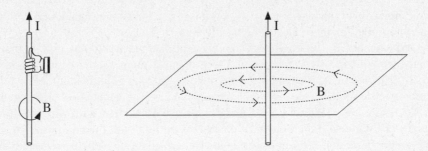

Fig. 3.1: Magnetic field around a straight conductor.

Oersted carried out a simple experiment to demonstrate this field around a current-carrying wire. A horizontal plane with a long wire passing through its centre has a number of compass needles placed on its surface.

When there is no current the needles align along the earth's magnetic field that is along the north-south direction. A current is now passed in the wire. The alignment of the needles changes showing that there is a magnetic field due to the current in the wire (Figure 3.2).

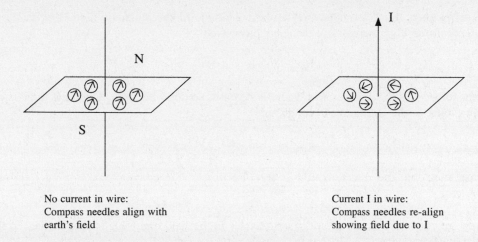

No current in wire:
Compass needles align with
earth's field

Current I in wire:
Compass needles re-align
showing field due to I

Fig. 3.2: Experiment to show field around a current in a long straight wire.

3.3 Force due to a Current-carrying Conductor placed in a Magnetic Field

A conductor in which a current I is flowing has an associated magnetic field as discussed in section 3.1.

Let the conductor be placed in a magnetic field B. The two fields — field due to the current and field B interact resulting in a force. This force now acts on the conductor. If the conductor is not massive, this force would move it.

Connect a high voltage source across two metal rails on which a movable conductor PQ of length ℓ is placed. The conductor PQ can move on these rails. A uniform magnetic field B cuts the assembly as shown in Figure 3.3. As current I flows through PQ, a force acts on the conductor and it starts moving to the right (bold arrow).

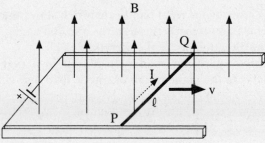

Fig. 3.3: The conductor PQ experiences a force and moves to the right.

By continuing with the experiment a number of observations can be made about the force:

1. If current is increased the conductor moves faster. The force increases as current increases and vice versa. It is found that the force is directly proportional to the current I:

$$\mathbf{F_{mag}} \qquad \alpha \qquad \mathbf{I} \qquad\qquad\qquad 3.4$$

2. It is experimentally determined that as the magnitude of the magnetic field B is increased, the magnetic force F_{mag} increases in the same proportion.

$$F_{mag} \qquad \alpha \qquad B \qquad\qquad 3.5$$

3. As the length ℓ of the conductor in the field increases so does the force. Again the force is directly proportional to the length of the conductor:

$$F_{mag} \qquad \alpha \quad \ell \qquad\qquad 3.6$$

Please note that the direction of ℓ is the direction of the conventional current I through the conductor.

4. When the conductor is placed at different angles to the field, the force is proportional to the sine of the angle between ℓ and **B**. Let the angle between ℓ and **B** be θ, then

$$F_{mag} \qquad \alpha \quad \sin\theta \qquad\qquad 3.7$$

Combining the equations 3.4, 3.5, 3.6 and 3.7 gives

$$F_{mag} \qquad \alpha \quad I \ell B \sin\theta \qquad\qquad 3.8$$

Since the direction of the force (Figure 3.4) is always perpendicular to the plane of ℓ and **B**, in vector form equation 3.8 can be written as a cross product:

$$F_{mag} \qquad = \quad I (\ell \times B) \qquad\qquad 3.9$$

The constant of proportionality is equal to 1 in the SI system of units.

The direction of this force is given by the **right hand rule which states that if the fingers of the right hand are curled from vector ℓ to vector B through the smaller angle, the force F will be in the direction of the thumb. The direction of ℓ is the direction of I.**
  We conclude that a current-carrying conductor placed in a magnetic field experiences a magnetic force perpendicular to the plane of the magnetic field and current.

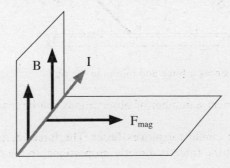

Fig. 3.4: F_{mag} is perpendicular to the plane of ℓ and **B**.

The magnetic force on a current-carrying conductor can vary from a minimum force of zero when $\theta = 0$

$$F_{mag} \quad = \quad I\ell B \ \sin 0°$$

$$F_{mag} \quad = \quad 0 \qquad\qquad 3.10$$

to a maximum of $I\ell B$ when $\theta = 90°$

$$F_{mag} \quad = \quad I\ell B \ \sin 90°$$

$$= \quad I\ell B \qquad\qquad 3.11$$

Using this maximum value of $\mathbf{F_{mag}}$ the **magnetic induction B** can be defined as

$$B \quad = \quad \frac{F}{I\ell} \qquad\qquad 3.12$$

The magnetic induction B is the force experienced by a conductor of unit length that is placed perpendicular to a magnetic field B in which a unit current is flowing.

In the SI system the unit for B is tesla (T).

$$B \quad = \quad \frac{F}{I\ell}$$

$$= \quad \frac{1N}{(1A)(1m)}$$

$$= \quad 1 \ T$$

The magnetic induction is one tesla (T) when it exerts a force of 1 newton (N) on a conductor of length 1 metre placed at right angles to the field in which a 1 ampere (A) current is flowing.

A smaller more practical unit is the gauss (G).

$$1 \ \text{tesla (T)} \quad = \quad 10^4 \ \text{gauss (G)} \qquad\qquad 3.13$$

It should be noted that a magnetic field directed into the plane of the paper is symbolized by

X   X   X   X

X   X   X   X

X   X   X   X

and the field out of the plane of the paper is depicted by

.   .   .   .

.   .   .   .

.   .   .   .

If two parallel conductors are placed, their fields interact and then it is observed that when currents are in the same direction, the forces (attractive forces) on the conductors are such that the conductors move towards each other. When currents are in opposite directions, they move away (repulsive forces) from each other (Figure 3.5).

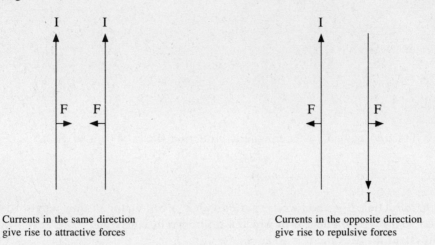

Currents in the same direction
give rise to attractive forces

Currents in the opposite direction
give rise to repulsive forces

Fig. 3.5: Parallel and anti-parallel currents and force directions.

Example: A wire in a magnetic field experiences a force of 0.5 N when the current in the wire is 3A. What is the current in the wire when it experiences a force of 0.075 N?

$$F_1 \quad = \quad 0.5N$$
$$I_1 \quad = \quad 3A$$
$$F_2 \quad = \quad 0.075N$$
$$I_2 \quad = \quad ?$$

The forces on the wire for the two currents are given by

$$F_1 \quad = \quad B\,I_1\ell\,\sin\theta$$
and
$$F_2 \quad = \quad B\,I_2\ell\,\sin\theta$$

Dividing F_2 by F_1 gives

$$\frac{F_2}{F_1} \quad = \quad \frac{I_2}{I_1}$$

$$I_2 \quad = \quad \frac{F_2\,I_1}{F_1}$$

$$= \quad \frac{(0.075\,\text{N})\,(3\,\text{A})}{0.5\,\text{N}}$$

$$= \quad 0.45\,\text{A}$$

3.4 Magnetic Flux and Flux Density

Similar to the electric flux φ, the magnetic flux Φ through a loop of area **A** placed in a magnetic field **B** is defined as

$$\Phi \quad = \quad \textbf{B. A} \qquad\qquad 3.14$$

$$= \quad B \, A \cos\theta \qquad\qquad 3.15$$

where **B** is the magnetic induction, **A** the area of the loop and θ the angle between **B** and **A** (Figure 3.6b). Note that area is a vector quantity and its direction is perpendicular to the plane surface.

When **B** is parallel to **A** (Figure 3.6a), $\theta = 0°$, $\cos\theta = 1$, Φ is maximum and equal to

$$\Phi \quad = \quad B \, A \qquad\qquad 3.16$$

When **B** is perpendicular to **A** (Figure 3.6c), $\theta = 90°$, $\cos\theta = 0$, Φ is minimum and equal to

$$\Phi \quad = \quad 0 \qquad\qquad 3.17$$

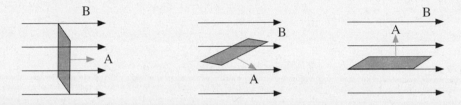

Fig. 3.6a: **B** parallel to **A**   *b:* **B** makes angle θ with **A**   *c:* **B** perpendicular to **A**

As in the case of electric flux, the magnetic flux is a scalar, and would vary with B, A and $\cos\theta$. **The unit of flux is a weber (Wb).** A weber is equal to **$1 \, T.m^2$.**

From equation 3.16, it can be seen that **the magnetic induction B can also be defined as the flux density. The flux density B is the magnetic flux per unit area:**

$$B \quad = \quad \frac{\phi \, \Phi}{A} \qquad\qquad 3.18$$

If flux is measured in webers and area in $metre^2$ then another unit for B in SI system of units is:

$$B \quad = \quad \frac{Wb}{m^2}$$

$$= \quad Wb \, m^{-2} \qquad\qquad 3.19$$

It may be noted that a $Wb \, m^{-2}$ is equal to a tesla (T).

3.5 Ampere's Law and Determination of Flux Density B

Ampere derived an expression known as Ampere's Circuital Law which gives the relation between the current in a closed loop and magnetic field produced by the wire.

Take an arbitrarily selected closed loop around a current as shown in Figure 3.7. The loop is divided into n small segments of lengths $\Delta\ell_1$, $\Delta\ell_2$, $\Delta\ell_3$,...., $\Delta\ell_n$. The magnetic field at each segment is B_1, B_2, B_3, B_n respectively. The product of each segment and the parallel component of the field B to each segment is given by the dot product of B and $\Delta\ell$ for each **B. $\Delta\ell$**

Ampere computed that the sum of all such products over the closed loop is equal to μ_o times the total current I bounded by the closed loop

$$\mathbf{B_1.\Delta\ell_1 + B_2.\Delta\ell_2 + B_3.\Delta\ell_3 +.........+ B_n.\Delta\ell_n} \quad = \quad \mu_o \, [I_1 + I_2 \,...] \qquad 3.20$$

$$\Sigma \, \mathbf{B_r.\Delta\ell_r} \quad = \quad \mu_o \, I_{enclosed} \qquad 3.21$$

μ_o **is a constant and is related to the medium. It is called the permeability of free space.** In free space its value is $4\pi \times 10^{-7}$ **Wb/A.m.**

Equation 3.21 is Ampere's Circuital Law

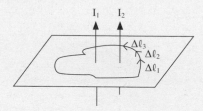

Fig. 3.7: Ampere's Circuital Law.

3.5.1 Field due to a Current-carrying Conductor

The magnetic flux due to a solenoid is to be determined. The field of a wire is intensified by coiling the wire into a solenoid. The solenoid is a long straight wire coiled into several closely spaced loops like a helical spring.

When a current is passed through a solenoid a magnetic field similar to that of a bar magnet is created (Figure 3.8). The advantage over a bar magnet is that by changing the current through the solenoid the field can be altered and secondly in a number of applications the field needs to be turned off and on. This can be done in a solenoid by switching on and off the current but a bar magnet is a permanent magnet.

A solenoid

A bar magnet

Fig. 3.8: Similar magnetic fields of a solenoid and a bar magnet.

The field inside the solenoid is strong and uniform. It is so weak outside that it can be considered as zero.

The field of a solenoid can be determined by applying Ampere's Circuital Law.

To understand this, let us cut across the solenoid through its centre and the two opposite edges are considered. It is seen that current goes into the plane of the paper in all loops at one edge and comes out of the plane of the paper at the opposite edge. Using the right hand rule to find the magnetic field lines around each loop gives the field pattern through the solenoid as shown in Figure 3.9. The field between the coils cancels out.

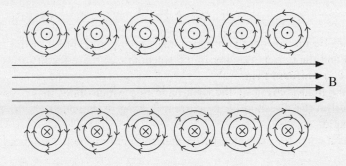

Fig. 3.9: Magnetic field due to a solenoid.

To apply Ampere's Circuital Law an arbitrary closed loop has to be selected. A rectangular loop PQRS of length ℓ and breadth b is drawn (Figure 3.10).

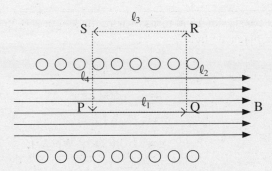

Fig. 3.10: An arbitrary loop PQRS is selected to apply Ampere's Circuital Law.

Applying Ampere's Circuital Law along rectangular loop PQRS

$$\Sigma\ \mathbf{B_r.\Delta\ell_r} \qquad\qquad = \mu_o\ I_{enclosed}$$

$$\mathbf{B_1.\Delta\ell_1} \quad + \mathbf{B_2.\Delta\ell_2} \quad + \mathbf{B_3.\Delta\ell_3} \quad + \mathbf{B_4.\Delta\ell_4} \quad = \mu_o\ I_{enclosed} \qquad 3.22$$

$$B_1\Delta\ell_1\cos0° + B_2\Delta\ell_2\cos90° \ + B_3\Delta\ell_3\cos0° \ + B_4\Delta\ell_4\cos90° = \mu_o\ I_{enclosed}$$

$$B_1\ell_1 \qquad\quad + 0 \qquad\qquad + B_3\Delta\ell_3\cos0° \ + 0 \qquad\quad = \mu_o\ I_{enclosed}$$

The field B_3 outside the solenoid is very weak and can be neglected.

$$B_1\ell_1\ +\ 0\ +\ 0\ +\ 0\ =\ \mu_o\ I_{enclosed} \qquad\qquad 3.23$$

If the number of turns of the solenoid per unit length is n, then the solenoid has $n\ell_1$ turns in length ℓ_1. Each turn carries a current I, therefore the current in $n\ell_1$ turns is

$$\text{Current in the loop PQRS} \quad = \quad n\ell_1\, I$$

Equation 3.23 becomes

$$B_1\ell_1 \quad = \quad \mu_o\,(n\ell_1\, I)$$

$$B_1 \quad = \quad \mu_o\, n\, I \qquad\qquad 3.24$$

The magnetic field inside the solenoid is proportional to the permeability of the medium, number of turns per unit length and the current flowing in the solenoid. All these factors are constants. **The field inside the solenoid is a uniform field**.

The direction of the field is given by the right hand rule which states that if you hold the solenoid so that the fingers curl in the direction of the current flowing in its coils then the thumb will point in the direction of the field.

If the solenoid has a core, for example of iron then the field inside the solenoid would be equal to

$$B \quad = \quad \mu_r\, \mu_o\, n\, I \qquad\qquad 3.24a$$

where μ_r is the relative permeability of the material of the core.

Example: A long solenoid consists of 1500 turns of wire. Its length is 0.75 m. A current of 5A flows in the solenoid. What is its magnetic field?

$$\ell \quad = \quad 0.75 \text{ m}$$
$$N \quad = \quad 1500 \text{ turns}$$
$$I \quad = \quad 5\text{A}$$
$$\mu_o \quad = \quad 4\pi \times 10^{-7} \text{ Wb/A.m}$$

The magnetic field in a solenoid is

$$B \quad = \quad \mu_o\, n\, I$$

$$= \quad \mu_o\, \frac{N}{\ell}\, I$$

$$= \quad \frac{(4\pi \times 10^{-7}\,\text{Wb/A.m})\,(1500)\,(5\text{ A})}{0.75 \text{ m}}$$

$$= \quad 1256.63 \times 10^{-5} \text{ T}$$

$$= \quad 1.256 \times 10^{-2} \text{ T}$$

3.6 Force on a Moving Charge in a Magnetic Field

A current comprises of moving charges. The force on a conductor is due to the forces on the moving charges in the conductor.

Let us assume that our conductor of length ℓ and area of cross-section A has a charge density equal to n. The charges are moving with a velocity $\mathbf{v}$ and they are covering a length ℓ in time Δt.

In terms of amount of charges the current I is equal to

$$I \quad = \quad \frac{\Delta Q}{\Delta t} \qquad\qquad 3.25$$

The amount of charge in length ℓ of the conductor which crosses the conductor in unit time is then calculated.

The number of charge carriers in the conductor can be calculated by:

$$\left(\frac{number}{unit\ volume}\right) (volume) \; = \; n\,(A\ell).$$

The amount of charge is

$$\Delta Q \quad = \quad \text{(Charge on each charge carrier) (Number of charge carriers)}$$

$$= \quad q\,(n\,A\ell)$$

Equation 3.25 becomes
$$I \quad = \quad \frac{qnA\ell}{\Delta t}$$

$$= \quad \frac{qnA\ell v}{\ell}$$

$$= \quad q\,n\,Av$$

Putting this value of I in equation 3.9

$$\mathbf{F_{mag}} \quad = \quad I\,(\boldsymbol{\ell} \times \mathbf{B})$$

$$= \quad qn\,Av\,(\boldsymbol{\ell} \times \mathbf{B})$$

In terms of magnitude and direction vector $\boldsymbol{\ell}$ is $\ell\hat{\ell}$. The direction of $\mathbf{v}$ is the direction of $\hat{\ell}$ as the charges move in direction of $\boldsymbol{\ell}$. Thus $v\,\hat{\ell}$ is equal to vector $\mathbf{v}$.

$$\mathbf{F_{mag}} \quad = \quad nqAv\,(\ell\hat{\ell} \times \mathbf{B})$$

$$= \quad nq\,A\ell\,(v\hat{\ell} \times \mathbf{B})$$

Force per charge $\qquad \dfrac{F_{mag}}{\#\ of\ charges} \quad = \quad \dfrac{nqA\ell\,(\mathbf{v} \times \mathbf{B})}{nA\ell}$

$$= \quad q\,(\mathbf{v} \times \mathbf{B}) \qquad\qquad 3.26$$

The direction of the force is given by the right hand rule.

The magnitude of the force $\quad\quad$ F_{mag} $\quad=\quad$ $I\ell B\ \sin\theta$ $\quad\quad\quad\quad\quad\quad\quad\quad\quad\quad$ 3.27

If the velocity **v** and magnetic field **B** are in the same direction $F_{mag} = 0$. If the angle between them is 90°, the force is maximum and equal to q v B.

$\quad$ Figures 3.11 give the magnetic force for different velocities and directions of the charge q. If the charge is stationary, there is **no force on it** (Figure 3.11a).

Fig. 3.11a: No force acts on a stationary charge q placed in a magnetic field **B**.

Next project a charge q with a velocity **v** in the field direction. The charge is not affected. It keeps moving at the same constant velocity **v**. **No force acts on it** (Figure 3.11b).

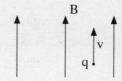

Fig. 3.11b: No force acts on a charge moving parallel to **B**.

Now project the charge q with a velocity **v** perpendicular to the field **B**. A force now acts on this charge and because of this force the charge is deflected into the plane of the paper (Figure.3.11c).

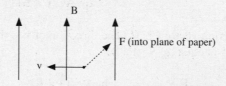

Fig. 3.11c: A charge projected with velocity **v** perpendicular to **B** experiences a force **F** directed into plane of paper.

This force is perpendicular to both **v** and **B** and would make the particle move in a circle of radius r. The centripetal force will be provided by this force (Figure 3.12).

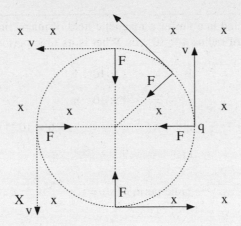

Fig. 3.12: Circular trajectory of a charged particle projected with velocity **v** into magnetic field **B**.

Therefore

$$\mathbf{F_{centripetal}} = \mathbf{F_{mag}} = q\,\mathbf{v} \times \mathbf{B}$$

$$\frac{mv^2}{r} = q\,v\,B\,\sin 90°$$

$$\frac{mv^2}{r} = q\,v\,B$$

$$r = \frac{mv}{qB} \qquad\qquad 3.28$$

The radius of the trajectory r is directly proportional to the mass and velocity and inversely to the charge and field.

Example: Calculate the radius of trajectory of a proton projected perpendicularly with velocity 5×10^5 m/s in uniform magnetic field of flux density 10^{-3} T.

$$B = 10^{-3}\text{ T}$$

$$v = 5 \times 10^5 \text{ m/s}$$

$$q = 1.6 \times 10^{-19} \text{ C}$$

$$m = 1.67 \times 10^{-27} \text{ kg}$$

$$r = \frac{mv}{qB}$$

$$= \frac{(1.67\times 10^{-27}\text{kg})\,(5\times 10^5\text{m/s})}{(1.6\times 10^{-19}\text{C})\,(10^{-3}\text{T})}$$

$$= 5.22 \text{ m}$$

Example: An electron projected in a uniform magnetic field of magnetic induction B 2×10^{-2} T moves in a circle of radius 4×10^{-2} m. What is its time period?

$$B \quad = \quad 2 \times 10^{-2} \text{ T}$$

$$r \quad = \quad 4 \times 10^{-2} \text{ m}$$

$$q \quad = \quad e \quad = \quad 1.6 \times 10^{-19} \text{ C}$$

$$m \quad = \quad 9.1 \times 10^{-31} \text{ kg}$$

$$\text{Time period T} \quad = \quad \frac{2\pi r}{v}$$

But

$$\frac{mv^2}{r} \quad = \quad B \, e \, v$$

$$v \quad = \quad \frac{B \, e \, r}{m}$$

Inserting the value of v in the equation for T

$$T \quad = \quad \frac{2\pi r m}{Ber}$$

$$= \quad \frac{2\pi m}{Be}$$

$$= \quad \frac{2\pi \, (9.1 \times 10^{-31} \text{kg})}{(2 \times 10^{-2} \text{T}) \, (1.6 \times 10^{-19} \text{m})}$$

$$= \quad 17.85 \times 10^{-10} \text{ s}$$

$$= \quad 1.785 \times 10^{-9} \text{ s}$$

$$= \quad 1.785 \text{ nanoseconds}$$

3.7 Motion of a charged particle in electric and magnetic fields

In the last chapter, we saw that a positively charged particle will accelerate in the direction of the electric field **E** and a negatively charged particle in the opposite direction (−**E**). If projected perpendicular to the electric field **E** the charged particle q will be deflected as shown in Figure 3.13.

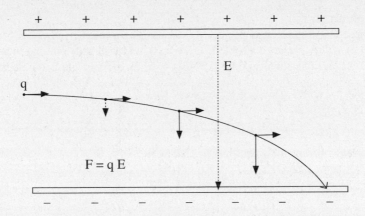

Fig. 3.13: Deflection of a charged particle q in an electric field of electric intensity **E**.

The positive charge will deflect towards the negative plate whilst the negative charge will deflect towards the positive plate. The electric intensity **E** is directed from the positive to the negative plate. As the particle is deflected there is a component of velocity in the direction of force F. This force does work on the charge and its velocity keeps on increasing in the direction of the force.

Thus the electric force $\mathbf{F_{el}} = q\ \mathbf{E}$ can accelerate and deflect a charged particle. If the particle is projected into a magnetic field with a velocity **v** it will experience a force perpendicular to both **v** and **B**. This force will deflect the particle only. This force is

$$\mathbf{F_{mag}} \quad = \quad q\ (\mathbf{v} \times \mathbf{B}).$$

Under the influence of both forces, the charged particle shall experience a force equal to the vector sum of $\mathbf{F_{el}}$ and $\mathbf{F_{mag}}$.

Therefore **the total force due to both forces which is called the Lorentz Force** is

$$\mathbf{F} \quad = \quad \mathbf{F_{el}} \quad + \quad \mathbf{F_{mag}}$$

$$= \quad q\ \mathbf{E}\ +\ q\ (\mathbf{v} \times \mathbf{B}) \qquad\qquad 3.29$$

3.8 Ratio e/m for an electron

The change to mass (e/m) ratio of charged particle is used to distinguish one charge from another. It is found that a number of charged particles have the same charge. For example, an electron and a muon have the same charge of -1.6×10^{-19} C, while a proton and a pion have the same positive charge. Thus to distinguish one particle from another the charge to mass ratio is determined. This ratio is specific to each charged particle as different particles might have the same charge but different masses or same mass but different charges.

3.8.1 e/m of an electron

An electron of charge e and mass m is projected into a uniform magnetic field B (direction out of the plane of the paper). It will experience a force **F** = e(**v** × **B**), and will start moving in a circular trajectory of radius r (Figure 3.14). (The force will be F = –q(v × b) as the particle is an electron).

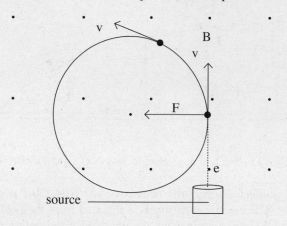

Fig. 3.14: Electron moves in a circular trajectory when projected perpendicular to the field.

The centripetal force is equal to the magnitude of the magnetic force

$$e \, v \, B \quad = \quad \frac{mv^2}{r}$$

$$\frac{e}{m} \quad = \quad \frac{v}{Br} \qquad\qquad 3.30$$

To determine e/m v, B and r have to be found.

The magnetic field B can be calculated if we know the current, radius and number of turns in the coil producing the magnetic field.

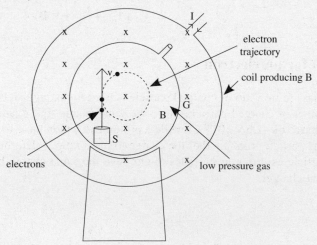

Fig. 3.15: Apparatus to determine e/m.

CHAPTER 3 Electromagnetism   **99**

The radius of the trajectory r can be found by making the beam visible. Air is evacuated from the spherical glass container G in which the experiment is being conducted. The container is now filled with gas at low pressure. Electrons are projected at high velocities into the magnetic field. They strike the gas atoms and cause excitation. When the excited electrons come back to their normal positions they emit light. If they are filled with hydrogen gas, blue light is emitted. Neon gas at low pressure would emit red light. Thus depending on the type of gas, the path of the electrons would become visible as a lighted circle and its diameter can be measured using a scale (Figure 3.15).

The velocity v can be found if the potential difference V through which the electrons are accelerated is known. The kinetic energy that the electrons gain when accelerated through this potential difference V is equal to Ve (Figure 3.16).

Therefore

$$KE = Ve$$

$$\tfrac{1}{2}\,m\,v^2 = V\,e$$

$$v^2 = \frac{2Ve}{m}$$

$$v = \sqrt{\frac{2Ve}{m}}$$

Equation 3.30 becomes

$$\frac{e}{m} = \frac{\sqrt{(2Ve)/m}}{Br}$$

Squaring

$$\frac{e^2}{m^2} = \frac{2Ve}{mB^2r^2}$$

$$\frac{e}{m} = \frac{2V}{B^2r^2} \qquad\qquad 3.31$$

Thus knowing V, B and r, the charge-to-mass e/m ratio of an electron can be determined.

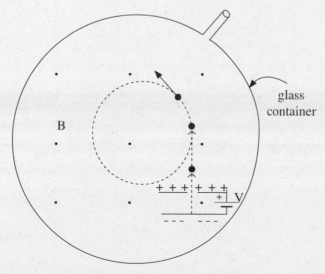

Fig. 3.16: Electrons are accelerated through potential difference V and enter the magnetic field with velocities v.

Particles of a particular velocity can also be selected by using crossed electric and magnetic fields. By adjusting the potential difference V the electric force $F_{electric}$ is made equal to the magnetic force $F_{magnetic}$. This is called velocity selector method. The charges that enter the chamber have a velocity given by:

$$F_{magnetic} = F_{electric}$$
$$qvB = qE$$
$$v = \frac{E}{B}$$

Putting this value of v in equation 3.30 gives another expression for e/m

$$\frac{e}{m} = \frac{E}{B^2 r}$$

If E, B and r are known e/m can be found.

3.9 Cathode Ray Oscilloscope (CRO)

The Cathode Ray Oscilloscope is a versatile and exciting application of the effect of electric field on charged particles. The cathode rays as the name suggests are rays coming from the negative electrode (cathode). They are a beam of fast electrons produced by heating the filament in an evacuated tube. The anode is cylindrical with a hole down its centre. It focuses the beam of electrons which move in straight lines through the hole in the anode. Another cylindrical anode accelerates this focused beam.

The rays next pass through two pairs of deflecting plates which deflect the beam vertically (y-plates) and horizontally (x-plates) before the beam strikes the fluorescent screen. The point where it hits the screen appears as a light spot (Figure 3.17).

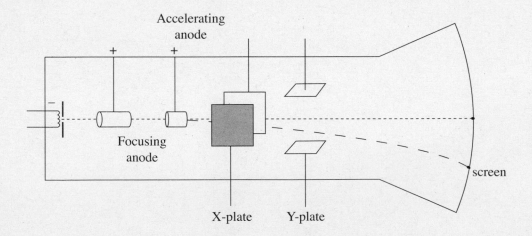

Fig. 3.17: The Cathode Ray Oscilloscope (CRO).

The x-plates are connected to time-base, saw-tooth voltage (Figure 3.18). The voltage rises linearly over a short interval of time and then falls quickly to zero. The spot on the screen moves on the screen at a constant speed from left to right then comes back very fast (Figure 3.19). Its coming back motion is so fast that it is not visible. If the time-base is very short the spot is moving so fast down the screen that it appears as a line across the screen. The x-plates only allow horizontal motion.

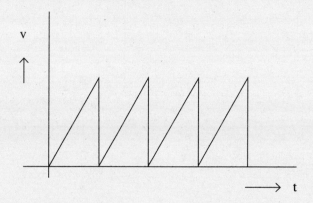

Fig. 3.18: Saw tooth voltage across the x-plates.

The external voltage that is being studied is applied to the y-plates. The y-plates deflect the spot in the vertical direction. Suppose a sinusoidal voltage has to be measured and visualized, it is applied to the y-plates. It will be moved up and down. At the same time, the x-plate time base will move it to the right and so a sinusoidal trace will appear moving down the screen.

The y-voltage can be observed on the screen provided the frequencies of the y-voltage and x-voltage match. Proper synchronization between the frequency of the y-plates and time-base frequency can make the trace stationary. In other words, the frequency of the time-base is adjusted so that the y-frequency f_y is a multiple of the x-frequency f_x.

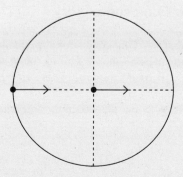

Fig. 3.19: The spot moves from left to right on the CRO screen.

If the external alternating frequency is made equal to the time-base frequency

$$f_x = f_y$$

then the stationary trace as seen in Figure 3.20 will be obtained.

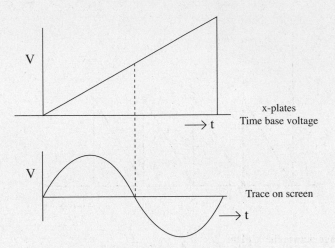

Fig. 3.20: The sine wave voltage across the y-plates as seen on the screen.

The CRO can be used as a voltmeter to determine an unknown voltage. The CRO is first calibrated by connecting the x- or y-plates to a known voltage. As the unknown voltage is switched on, the spot moves away from the centre position. This point is noted. By comparing the positions of the spot with the calibrated value the unknown voltage is determined.

Using CRO, waveforms can be displayed, studied and compared. Their phases, differences in phase and frequencies can also be determined.

3.10 Torque on a current-carrying conductor

In section 3.3, it was seen that a current-carrying conductor experiences a force when placed in magnetic field (Equation 3.9). If a current-carrying coil that is free to rotate is suspended in the field, the force on it will produce a torque and it will rotate the coil. This principle has been applied in designing galvanometers and electric motors.

Let a rectangular single coil ABCD be placed in a uniform magnetic field **B** as shown in Figure 3.21.

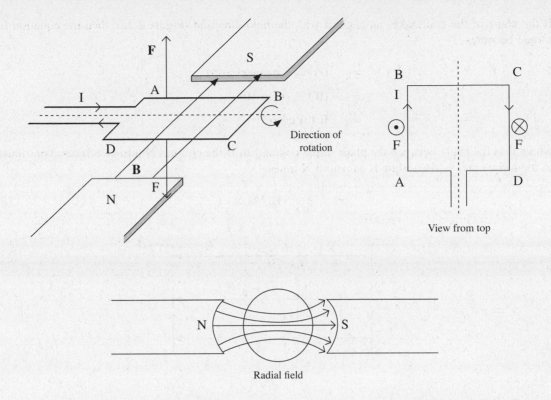

Fig. 3.21: A rectangular conductor ABCD in which a current I is flowing is placed in a magnetic field B.

A current I flows through this coil. Let us consider the arm AB. Current in this arm moves to the right, the magnetic field B is directed into the plane of the paper. It is perpendicular to current I and so the force on this arm AB is

$$\mathbf{F_{mag}} \quad = \quad I\,\ell \times \mathbf{B}$$

The force direction can be found by applying the right hand rule. On the arm AB it is directed upwards. The magnitude of the force is

$$F_{mag} \quad = \quad I\,\ell\,B \sin 90°$$
$$= \quad I\,\ell\,B$$

Force $I\,\ell\,B$ acts on arm CD as well but it is directed downwards as current in this arm is flowing in the opposite direction to the current in AB. On sides BC and DA there is no force as I and B are parallel. The two forces act as a couple and can rotate the coil about its axis.

The torque or moment of force which is responsible for this angular motion is equal to

$$\text{Torque} \quad \tau \quad = \quad \text{(Force) (moment arm)}$$
$$= \quad (I\,\ell\,B)\,(b) \qquad \ell B = \text{area of coil}$$
$$= \quad B\,I\,A \qquad\qquad\qquad\qquad 3.32$$

If the plane of the coil makes an angle α with the field direction (Figure 3.22), then the equation for torque becomes

$$\tau \quad = \quad \text{(Force) (moment arm)}$$

$$= \quad (B\ I\ \ell)\ (b\ \cos\alpha)$$

$$= \quad B\ I\ A\ \cos\alpha$$

where α is the angle between the plane of the coil and B. If the coil has N turns, each turn contributes to the torque and so the torque is enhanced N times.

$$\tau \quad = \quad BIAN\ \cos\alpha \qquad\qquad\qquad 3.33$$

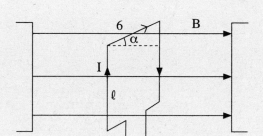

Fig. 3.22: The field makes an angle with the current direction in arms BC and DA.

The torque depends on the magnetic induction B, current I, area A of conductor, number of turns N and the orientation of the coil with respect to the magnetic field.

The equation for torque is valid for circular coils and radial fields. In radial fields I and B will be perpendicular.

When the plane of the coil is perpendicular to the field, torque is maximum:

$$\tau_{max} \quad = \quad BIAN\ \cos 0°$$

$$= \quad BIAN \qquad\qquad\qquad 3.34$$

When the plane of the coil and field are in the same direction, torque is minimum and equal to zero:

$$\tau_{min} \quad = \quad B1AN\ \cos 90°$$

$$= \quad 0 \qquad\qquad\qquad 3.35$$

The SI unit of torque is N.m.

Example: A coil of area 5×10^{-4} m^2 and 150 turns is placed in a magnetic field of magnitude 0.5 T. If a current 0.03A flows through the coil, what is the maximum torque?

$$B \quad = \quad 0.5 \text{ T}$$

$$N \quad = \quad 150 \text{ turns}$$

$$A \quad = \quad 5 \times 10^{-4} \text{ m}^2$$

$$I \quad = \quad 0.03 \text{ A}$$

Applying equation 3.33

$$\tau_{max} \quad = \quad BIAN \cos 0°$$

$$= \quad BIAN$$

$$= \quad (0.5 \text{ T}) (0.03\text{A}) (5 \times 10^{-4} \text{ m}^2) (150 \text{ turns})$$

$$= \quad 11.25 \times 10^{-4} \text{ Nm}$$

Example: A length of wire 0.1m is made into a circular loop and placed in a 1.2 T magnetic field. If the current in the wire is 1.5 A, what is the maximum torque the wire loop experiences?

$$B \quad = \quad 1.2 \text{ T}$$

$$I \quad = \quad 1.5 \text{ A}$$

$$\ell \quad = \quad 2 \pi r \quad = \quad 0.1\text{m}$$

Torque is given by

$$\tau \quad = \quad BIAN \cos \alpha$$

$$\tau_{max} \quad = \quad BIAN$$

For 1 turn

$$\tau \quad = \quad BIA$$

To find area, radius r has to be found.

Length of wire is drawn into a circular loop

$$2 \pi r \quad = \quad 0.1\text{m}$$

$$r \quad = \quad \frac{0.1}{2\pi}$$

$$A \quad = \quad \pi r^2$$

$$= \quad \frac{\pi (0.1)^2}{4\pi^2}$$

$$= \quad 7.96 \times 10^{-4} \text{ m}^2$$

The maximum torque

$$\tau \quad = \quad (1.2 \text{ T})(1.5 \text{ A}) (7.96 \times 10^{-4} \text{ m}^2)$$

$$= \quad 14.328 \times 10^{-4}$$

$$= \quad 1.433 \times 10^{-3} \text{ N m}$$

3.11 Galvanometer

A galvanometer is a simple instrument in which a coil is suspended by a strip along its axis of rotation in a magnetic field. The strip has a pointer which can move over a scale. As the current passes through the coil, it rotates and causes the strip to turn as well. The pointer moves on the scale showing whether current is small or large depending on the size of the deflection.

Attached to the strip is a spring. As the strip rotates, the spring tightens. When current is switched off, the spring uncoils and brings the strip, pointer and coil to its normal position.

When a current flows in the coil due to the effect of the field due to the current and magnetic field B, a torque is produced. This torque rotates the coil as far as it can against the opposing torque of the spring. As the current rises, the torque also increases and the coil turns more. The galvanometer deflection is thus a measure of the current.

The torque producing the deflection is

$$\tau_{deflecting} \quad = \quad BIAN \cos\alpha$$

If the magnetic field is made radial in all positions of the coil, the field direction and plane of coil remain parallel, then

$$\tau_{deflecting} \quad = \quad BIAN \cos 0°$$

$$= \quad BIAN$$

When the coil turns the spring, winding starts to tighten. According to Hooke's Law, the restoring force is proportional to the displacement. For angular motion, the restoring torque is proportional to the angular displacement.

$$\tau_{restoring} \quad \alpha \quad c\,\theta$$

At equilibrium
$$\tau_{deflecting} \quad = \quad \tau_{restoring}$$
$$BIAN \quad = \quad c\,\theta$$
$$I \quad = \quad \frac{c\theta}{BAN} \qquad\qquad 3.35$$

B, A, N and c are all constants and so

$$I \quad \alpha \quad \theta \qquad\qquad 3.36$$

The current through the galvanometer is directly proportional to the deflection angle.

3.11.1 Measuring the deflection angle θ of a galvanometer

You have used galvanometers in your physics laboratory. They are the **pivotal type**. A horse shoe magnet provides the magnetic field **B**. A small coil is suspended between the poles of the magnet and in the field B. The coil is wound on an iron frame to increase the torque when current flows through the coil. The pointer of a light weight material is attached to the coil which is kept in place by two bearings. The bearings allow the coil to rotate.

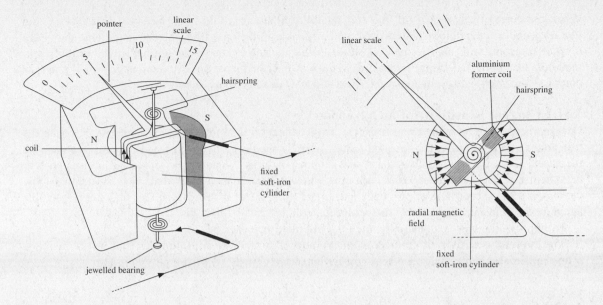

Fig. 3.23a: The moving-coil galvanometer.   *b:* Top-view of the moving galvanometer.

Two hair-like springs are attached to the coil on both sides which also serve as current leads. These springs bring the pointer back to the zero position after current is switched off. As the current flows through the coil of the galvanometer, the coil deflects along with the pointer. The pointer moves on a calibrated scale (Figure 3.23).

Good quality pivotal type galvanometers can be constructed to give deflections to current around 0.1 microamperes.

Another method of measuring currents is by what is termed **a lamp and scale** arrangement instead of a pointer. A very light round mirror is attached to the suspension strip. Light from a lamp is beamed on to the mirror. The reflected light falls on an opaque metre scale and is visible as a spot on the scale. As the galvanometer coil deflects, the mirror also turns and the light spot moves on the scale. The amount of deflection can be measured on the metre scale (Figure 3.24). The sensitivity of these galvanometers is 1000 times more than that of the pivotal type.

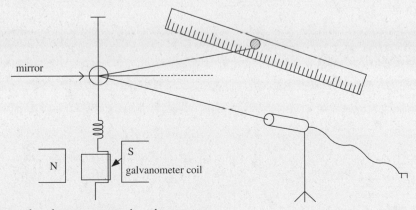

Fig. 3.24: Lamp and scale arrangement in galvanometers.

When the current is switched off, the coil should oscillate as would happen in any vibratory system like a spring or a pendulum. This would not be suitable in the case of a galvanometer. One may have to wait for some time before the coil stops oscillating and is ready for the next reading. Different methods are applied to bring the pointer to zero at once without any oscillations. This is known as damping and such galvanometers are termed stable or dead beat galvanometers.

3.11.2 Current Sensitivity of a Galvanometer

Galvanometers can be made more sensitive if the deflection is more for a small current. To make the instrument more sensitive the constant value c/(BAN) should be decreased or B, A and N should be increased and c decreased. The area A and number of turns N cannot be increased as the coil will become heavy and the light metallic strip from which it is suspended will break. The constant c can be decreased if the suspension strip is made longer and its diameter is decreased. This may not be possible again as the strip may not be able to support the coil.

To improve the current sensitivity, the magnetic field is increased.

The current sensitivity is defined as the amount of current in milliamperes that would produce a one millimeter deflection on a scale placed one metre away from the galvanometer.

3.12 Measuring Currents and Voltages

Ammeters and voltmeters are instruments designed to measure currents and potential differences respectively. Ammeters, voltmeters and ohmmeters are modified galvanometers. They can be digital or analog.

3.12.1 Ammeter

A galvanometer is a sensitive instrument whose working is based on magnetic forces. It can detect small currents.

Suppose a galvanometer can detect small currents from 0 to I_g. We wish to construct an ammeter using this galvanometer so that the ammeter can measure currents up to a current equal to I. Current I is much greater than I_g.

When current I is in the circuit, current I_g should pass through the galvanometer to give full scale deflection, while the rest of the current $(I - I_g)$ should flow through a shunt resistance placed parallel to galvanometer coil as in Figure 3.25. The shunt should be a low resistance R_s so that $(I - I_g)$ current passes through it.

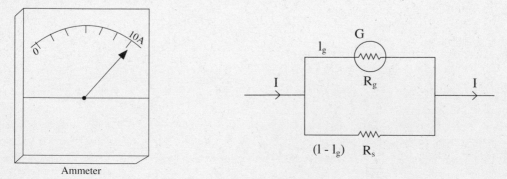

Fig. 3.25: Circuit for conversion of a galvanometer in an ammeter.

The shunt resistance R_s to be connected across the galvanometer of resistance R_g to convert it into an ammeter of range 0 to I can be determined easily. Since the galvanometer of resistance R_g and shunt resistance R_s are connected in parallel, the potential difference across them is equal.

The potential difference V is

$$V \quad = \quad I_g R_g \quad = \quad (I - I_g) R_s$$

$$R_s \quad = \quad \frac{I_g R_g}{I - I_g} \qquad\qquad 3.37$$

The galvanometer scale is calibrated so that it reads currents from $0 - I$.

Example: The full scale deflection for a galvanometer is 5 mA. Its resistance is 50 Ω. Find the shunt resistance that would convert the galvanometer into an ammeter with full scale deflection of 5 A?

$$I_g \quad = \quad 5 \text{ m A}$$

$$R_g \quad = \quad 50 \ \Omega$$

$$I_s \quad = \quad I - I_g$$

$$= \quad 5 \text{ A} - 5 \text{ m A}$$

$$= \quad 4.995 \text{ A}$$

$$R_s \quad = \quad ?$$

$$R_s \quad = \quad \frac{I_g R_g}{I - I_g} \quad = \quad \frac{(5 \times 10^{-3} \text{A}) (50 \Omega)}{4.995 \text{ A}}$$

$$= \quad \frac{0.25}{4.995}$$

$$= \quad 0.05005 \ \Omega$$

3.12.2 Voltmeter

To convert a galvanometer of resistance R_g into a voltmeter, a high resistance R_x is connected in series with the galvanometer. The galvanometer would show full scale deflection for a small current I_g and correspondingly low voltage $V_g = I_g R_g$ (V= IR). If the galvanometer is to measure higher voltages up to V, a high resistance R_x must be connected in series ensuring that voltage $(V - V_g)$ drops across R_x and voltage V_g drops across R_g (Figure 3.26).

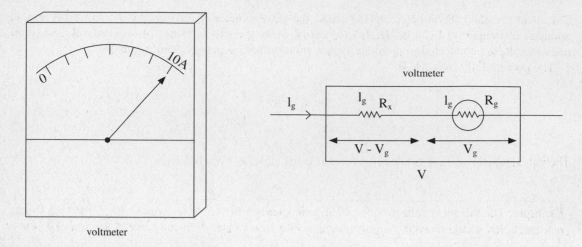

Fig. 3.26: Circuit for conversion of a galvanometer into a voltmeter.

If the current through R_x and R_g is I_g then V, the maximum voltage to be measured, is the sum of the voltages V_g and V_R:

$$V = V_g + V_R$$

$$V = I_g R_g + I_g R_x$$

$$R_x = \frac{V - I_g R_g}{I_g}$$

$$= \frac{V}{I_g} - R_g \qquad 3.38$$

Example: What resistance is to be connected in series to convert a 50 Ω galvanometer which shows full scale deflection for 5 mA into a voltmeter to measure 10 V maximum?

$$R_g = 50\ \Omega$$

$$I_g = 5\ mA$$

$$V = 10\ V$$

$$R_x = ?$$

Using equation 3.38

$$V = V_g + V_g$$

$$V = I_g R_g + I_g R_x$$

$$10\ V = (5\ mA)\ (50\ \Omega) + (5mA)(R_x)$$

$$(5mA)\ R_x = 10\ V - 0.25\ m\ V$$

$$R_x = \frac{9.75}{0.005}$$

$$= 1950\ \Omega$$

Example: (contd.)

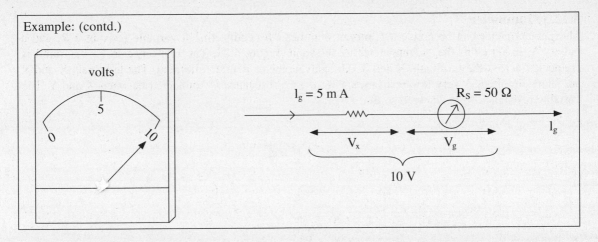

In general, the ammeter resistance must be very small otherwise the current in the circuit will change.

Ammeters and voltmeters circuits can be designed to make multi-range ammeters and voltmeters.

For instance, two or more shunt resistances of different values can be connected through keys across the galvanometer (Figure 3.27). Each shunt value is calculated so that the ammeter can measure different currents. The one in the diagram has three ranges $0 - 0.1A$, $0 - 1A$ and $0 - 10A$. For each range a different R_s is connected.

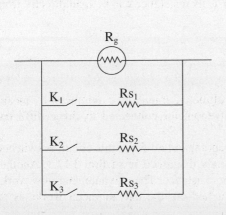

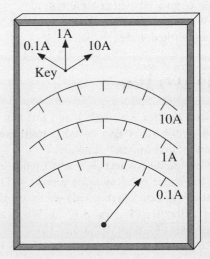

Fig. 3.27: A multi-range ammeter.

It should be remembered that on the galvanometer scale, different calibrated scales for current and voltage are drawn.

When K_1 is switched on, Rs_1 is in the circuit and the lowest ammeter scale $0 - 0.1$ A should be used. Similarly, when K_2 is switched on, Rs_2 is in circuit and the ammeter is now calibrated to measure currents from 0 to 1 A. When K_3 is switched on, Rs_3 is in the circuit. This is a resistance calculated to convert the galvanometer into a $0 - 10$ A ammeter. The keys K_1, K_2 and K_3 can be switched on from the external key.

3.12.3 Ohmmeter

The galvanometer can be converted into an ohmmeter by connecting a variable resistance R_s and a battery V in series to the galvanometer as shown in Figure 3.28. The resistance to be measured is connected across the terminals X and Y. The galvanometer is first calibrated. The terminals X and Y are short circuited. A very low resistance wire of zero resistance is connected between X and Y. The variable resistance R_s is changed so that

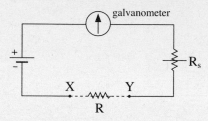

Fig. 3.28: An ohmmeter.

the galvanometer shows full scale deflection. The galvanometer scale is calibrated as zero. The short circuiting wire is now removed. This gap in the circuit means an infinite resistance. This position is labeled ∞ resistance. Other known resistances are now connected between X and Y and the galvanometer scale is now labeled completely.

When an unknown resistance is connected between X and Y, its resistance can be read directly from the calibrated scale.

3.13 AVO Meter

An AVO meter is a single meter that can be changed into a voltmeter, ammeter or ohmmeter by means of a switch. It is also called a **multimeter.** It uses one galvanometer connected to **three different circuits**, one for each function.

One circuit connects a low resistance in parallel to the galvanometer and when this circuit is switched on, the meter behaves as an ammeter. This is the circuit already discussed in section 3.12.1. Another circuit connects a high resistance in series to the galvanometer resistance. The galvanometer now works as a voltmeter (Section 3.12.2). When a third circuit is switched on, the galvanometer works as an ohmmeter (Section 3.12.3).

In a similar way, the AVO meter can be made to measure different ranges of voltages, currents and resistances as already mentioned in section 3.12.2 for ammeters.

Summary

The magnetic field at distance r from a long, straight wire carrying current I has the magnitude

$$B = \frac{\mu_o I}{2\pi r}$$

where $\mu_o = 4\pi \times 10^{-7}$ T.m/A is the **permeability of free space.** The magnetic field lines around a long, straight wire are circles concentric with the wire.

The **magnetic force** that acts on a charge q moving with velocity **v** in a magnetic field **B** has the magnitude

$$F = qvB \sin \theta$$

where θ is the angle between **v** and **B.**

To find the direction of this force, the **right-hand rule is applied:** Place the fingers of your open right hand in the direction of **B** and point your thumb in the direction of the velocity **v.** The force **F** on a positive charge is directed out of the palm of your hand. If the charge is negative rather than positive, the force is directed opposite the force given by the right-hand rule.

The SI unit of magnetic field is the **tesla** (T), or webers per square meter (Wb/m^2). Another unit commonly used is the Gauss (G); $1T = 10^4 G$.

If a straight conductor of length ℓ carries current I, the magnetic force on that conductor when it is placed in a uniform external magnetic field **B** is

$$F = BI\ell \sin \theta$$

The right-hand rule also gives the direction of the magnetic force on the conductor. In this rule, the thumb is in the direction of the current I.

The torque τ on a current-carrying loop of wire in a magnetic field **B,** has the magnitude

$$\tau = BIAN \cos\alpha$$

where I is the current in the loop and A is its cross-section area. The angle between **B** and a line drawn perpendicularly to the plane of the loop is α.

If a charged particle moves in a uniform magnetic field B, its initial velocity is perpendicular to the field and it will move in a circular path whose plane is perpendicular to the magnetic field. The radius r of the circular path is

$$r = \frac{mv}{qB}$$

where m is the mass of the particle and q is its charge.

Ampere's Law can be used to find the magnetic field around certain simple current-carrying conductors. It can be written as

$$\sum_{r=1}^{n} B\mathring{r}\Delta\ell = \mu_o I$$

$$\sum B_{\parallel}\Delta\ell = \mu_o I$$

where $B_{\parallel}$ is the component of B parallel to a small current element of length $\Delta\ell$ that is part of a closed path, and I is the total current that is enclosed in the closed path.

The **magnetic field inside a solenoid** has the magnitude

$$B = \mu_o n I$$

where n is the number of turns of wire per unit length, $n = N/\ell$

The **force between two parallel current-carrying conductors** is attractive if the currents are in the same direction and repulsive if they are in opposite directions.

The current through the galvanometer is directly proportional to the deflection angle.

A **galvanometer can be converted to an ammeter** by connecting a shunt resistance in parallel to the galvanometer. The value of the shunt is

$$R_s = \frac{I_g R_g}{I - I_g}$$

A **galvanometer can be converted into a voltmeter** by connecting a high resistance in series with the galvanometer. Its value is given by

$$R_x = \frac{V}{I_g} - R_g$$

The **galvanometer can be made into an ohmmeter** by connecting a variable resistance R_s and a battery V in series to the galvanometer.

An **AVO meter** is a single meter that can be changed into a voltmeter, ammeter or ohmmeter. It is also called a **multi-meter.** It uses one galvanometer connected to **three different circuits**, one for each function.

Questions

1. A charged particle is moving in a magnetic field. Which quantity is never affected by the field: particle's mass, velocity, linear momentum or kinetic energy?
2. Can a constant magnetic field make an electron at rest move? Explain your answer.
3. Why must you be careful when an ammeter is connected to a circuit?
4. Explain how the internal circuit of a volt-meter differs from that of an ammeter. Why are these circuits different?
5. A current-carrying conductor experiences no magnetic force when placed in a certain position in a uniform magnetic field. Explain.
6. What do you understand by the term time base of a CRO?
7. What gives rise to the couple that opposes the deflecting couple of the galvanometer?
8. How can a galvanometer be made more sensitive?
9. Distinguish between electric generators and electric motors?
10. A negative charge is moving west when it enters a magnetic field acting vertically downwards. In what direction is the force on the charge?

11. In what ways are the electric and magnetic fields different?
12. What is the direction of the force due to these currents in the diagram given below? Please show by arrows.

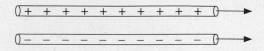

13. What is the direction of the force on a proton moving through the magnetic field **B** with velocity **v**.

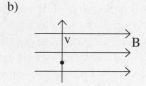

a)

b)

14. If you know the direction of the current in a coil, how can you determine which end is a magnetic 'S' pole?

Problems

1. Find the magnetic induction at a distance of 10 cm from a straight wire through which a 50 A current is flowing. Show by diagram the direction of this field.
2. What is the flux density in air at a point 0.05 m from a long straight wire carrying a current of 10 A?
3. A proton moves at 7.5×10^6 m/s along the x-axis. It enters a magnetic field of 2.8 T directed at an angle of 50° with the x-axis.

Calculate the force on and acceleration of the proton.

4. A wire carrying a current of 100 A due east is suspended between towers 50 m apart. The earth's magnetic field is in a northerly direction and its magnitude is 0.3 G. Find the force on the wire exerted by the earth's field.

5. An automobile has a speed of 30 m/s on a road where the vertical component of the earth's field is 8×10^{-5} T. What is the potential difference between the end of its axles which are 2 m long?

6. A current of 20 mA is maintained in single circular loop with a circumference of 2 m. A magnetic field of I T is directed parallel to the plane of the loop. What is the magnitude of the torque exerted by the magnetic field on the loop?

7. A circular wire loop of radius 50 cm is placed perpendicularly in a 0.6 T magnetic field. The current in the loop is 2.5 A. What is the magnitude of the torque?

8. Find the radius of curvature of a 50 eV electron in a magnetic field of 0.02 T.

9. Two singly ionized atoms move in semicircles into a magnetic field (0.1 T) of a mass spectrometer. Their speed is 1×10^{6} m/s. One ion is a proton of mass 1.67×10^{-27} kg and the other a proton of mass 3.34×10^{-27} kg. Find their distance of separation.

10. The radius of the path of an electron in a uniform magnetic field is 1×10^{-2} m. Find the time period of the orbit?

11. A thin horizontal copper rod is 1m long and has a mass of 50 g. What is the minimum current in the rod that can cause it to float in a horizontal magnetic field of 2 T?

12. A singly charged positive ion has a mass of 2.5×10^{-26} kg. After being accelerated through a potential difference of 250 V, the ion enters a magnetic field of 0.5 T in a direction perpendicular to the field. Calculate the radius of the ion's path in the field.

13. A wire carries a current of 7 A along the x-axis and another wire carries a 6 A current along the y-axis. What is the magnetic field at point **P** located at x = 4 m, y = 3m?

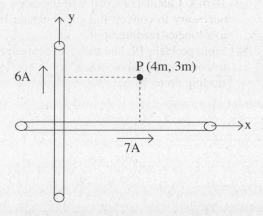

14. A solenoid has 100 turns of wire and has a length of 15 cm. It carries a current of 1 A. Find the magnetic field inside the solenoid.

15. A charged particle is projected perpendicularly into a magnetic field at a speed of 1400 m/s and experiences a force of magnitude F. If the speed of the particle were 1800 m/s, at what angle θ (less than 90°) with respect to the field would the particle experience the same force magnitude F?

16. The sensitivity of a moving coil galvanometer can be increased by
    a) making the coil of small cross-sectional area
    b) making a coil with less number of turns
    c) using springs with smaller force constant
    d) removing iron core from coil

17. The flux of 8×10^{-6} Wb is produced in the iron core of a solenoid. When the iron core is removed, a flux of 4×10^{-9} Wb is produced in the same solenoid by the same current. What is the relative permeability of iron?

18. A moving coil galvanometer gives full-scale deflection 0.005 A. It is converted into a voltmeter reading up to 5 V using an external resistance of 975 Ω. What is the resistance of the galvanometer?

19. A galvanometer of internal resistance 100 Ω gives a full-scale deflection for a current of 10 mA. Calculate the value of the resistance necessary to convert the galvanometer into a voltmeter reading up to 5 V.

20. From problem 19, find the resistance needed to convert the galvanometer to an ammeter reading up to 10 A.

21. A 30 Ω galvanometer deflects full scale when a current of 5×10^{-3} A is applied. Determine the resistance and type of connection needed to convert this galvanometer into an ammeter reading 0.5 A full scale.

22. A 30 Ω resistance can convert a galvanometer of 20 Ω into a voltmeter of 90 V. Find the value of the resistance which is required for a voltmeter of 30 V.

4 Electromagnetic Induction

OBJECTIVES

After studying this chapter, students should be able:
- to understand that a changing magnetic flux through a closed circuit induces an emf in it.
- to understand that the induced emf is only generated during the time that the flux is changing.
- to understand the motional emf.
- to have a clear concept of Faraday's Law.
- to understand self and mutual inductances.
- to understand that the unit for self inductance is a henry.
- to understand that energy is stored in the magnetic field of an inductor.
- to understand the working of AC and DC generators and DC motor.
- to have an idea of back emf and back motor effect and how they effect currents in circuits.
- to understand the transformer principles and applications.
- to have an idea about power losses in transformers.

4.1 Introduction

In the last chapter, we saw that an electric current produces a magnetic field. Can a magnetic field produce an electric current? A number of scientists tried to prove the phenomena but were not successful. **Electromagnetic induction**, the focus of this chapter, was discovered independently by Michael Faraday in England and Joseph Henry in USA in 1831.

It was found that the magnetic field can produce an induced emf, which in turn produces an induced current. The process of producing an emf with the help of magnetic field is termed **electromagnetic induction**.

Everyday important applications of this phenomenon are the production of electricity at power plants and its transportation to our homes via transformers.

4.2 Induced emf and Induced current

When a loop of wire is kept stationary in a magnetic field, no current flows through it. Faraday found that if this loop is moved across a magnetic field, it will cut the field and then current is detected in the wire. This current can be detected by a galvanometer attached to the wire loop. It is called an **induced current** (Figure 4.1).

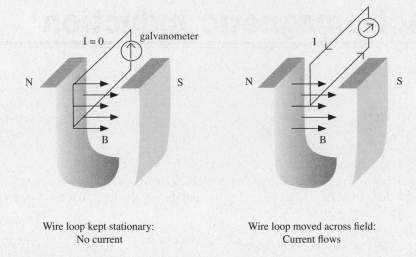

Fig. 4.1: An induced emf is produced when a conductor placed in a magnetic field is moved.

If the loop is pushed in the opposite direction, the current direction also reverses.

It was also noted that if the magnet is moved towards a stationary loop, induced current is produced. If the magnet is moved away (Figure 4.2a and b) from the loop, the current direction is in the opposite direction.

Fig. 4.2a: The magnet is moved left to right; induced current produced.

Fig. 4.2b: Magnet moved right to left; induced current reverses.

Experiments show that if the loop/magnet is moved faster, that is with higher velocity, current is more. If moved in a stronger magnetic field the current increases.

We can summarize that:

1. Moving a loop across a magnetic field that cuts across the loop produces an induced emf and induced current in the loop.
2. When the direction of motion is reversed, the current flows in the opposite direction.
3. The faster the loop is moved, the greater the current.
4. The current is more when field is increased.

One would imagine that the induced emf is due to motion. That is not true.

A solenoid coil is connected to a battery and a rheostat. A magnetic field is produced in the solenoid. Place a closed test loop within this field. **No** current is observed (Figure 4.3a).

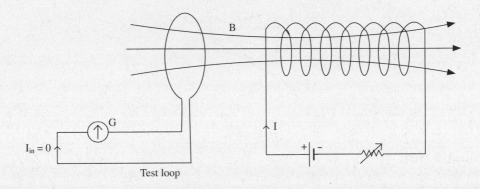

Fig. 4.3a: Current I kept constant: No induced current in test loop.

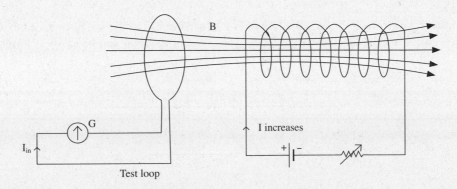

Fig. 4.3b: Current I is increased: Galvanometer registers induced I_{in}.

As the current in the coil is increased, the magnetic field increases and it is observed that an induced current results in the test loop (Figure 4.3b). When the current is decreased, the field decreases and the induced current reverses.

Thus, it is the changing magnetic field through the coil which produces the induced current. An emf is needed to produce current. The changing magnetic field produces an induced emf in the coil which produces the induced current. **The induced emf and induced current are only produced during the time the magnetic field change occurs.**

The direction of the induced current I_{in} is given by the **right hand rule (palm rule).**

As shown in Figure 4.4, when the palm of hand is stretched, the fingers of the right hand point in the direction of the magnetic field **B** and the outstretched thumb in the direction of **v**, the palm then points in the direction of the induced current (conventional current).

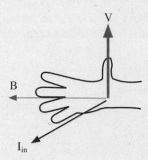

Fig. 4.4: The Right Hand Rule.

4.3 Motional emf

The emf produced in a conductor moving across a magnetic field is the motional electromotive force.

Consider a conductor AB of length ℓ which can move on a U-shaped metallic conductor placed perpendicular to the magnetic field **B** (Figure 4.5). Let us now move the conductor PQ with a velocity **v** to the right.

As AB moves with velocity **v**, an induced emf is produced across its ends. This induced emf is called the motional **emf ε, or the electromotive force**. The value of this emf can be easily determined.

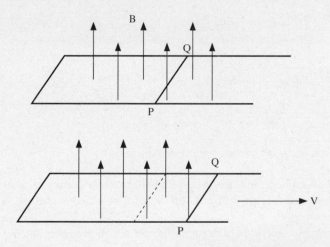

Fig. 4.5: Conductor AB moves to the right with velocity **v**.

The conductor has charges. The magnetic force on a charge q in a wire when the wire moves with a velocity **v** in a magnetic field **B** is given by

$$\mathbf{F_{magnetic}} \quad = \quad q\,(\mathbf{v} \times \mathbf{B})$$

Its magnitude is $\qquad F_{magnetic} = q\,v\,B \qquad$ when **v** and **B** are perpendicular.

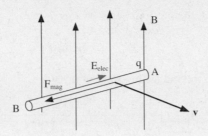

Fig. 4.6: The force **F** on charge q when conductor is moved to the right with velocity **v**. Also showing direction of resulting electric field **E**.

The direction of **F**$_{magnetic}$ would be as shown in the above Figure 4.6 using the right hand rule. The charge q moves from end A of the conductor AB to the other end B under the influence of this force. During the time in which AB is kept moving with velocity **v,** this force continues to push charges towards the end B. Thus, positive charges start accumulating at B, whilst the end A becomes negatively charged.

The potential difference between the two ends due to the accumulation of charges is the **induced emf or motional emf** ε. An electric field **E** directed from B to A arises. The accumulation of charges continues till a state of equilibrium is attained—electric force **E**$_{electric}$ becomes equal to the oppositely directed magnetic force **F**$_{magnetic}$

$$\mathbf{F}_{electric} = \mathbf{F}_{magnetic}$$
$$q\mathbf{E} = q\,(\mathbf{v} \times \mathbf{B}) \qquad\qquad 4.1$$

In terms of magnitude

$$qE = qvB$$
$$E = vB \qquad\qquad 4.2$$

The induced emf ε is the potential difference across AB. The electric intensity is

$$E = \frac{-\varepsilon}{\ell} \qquad\qquad \text{as } E = -V/d$$

Inserting this value of E in equation 4.2 and rearranging

$$\frac{\varepsilon}{\ell} = -v\,B$$

The motional emf ε, **is given by**

$$\varepsilon = -v\,\mathbf{B}\,\ell \qquad\qquad 4.3$$

where **v, B** and ℓ are mutually perpendicular.

It can be seen that when v = 0, ε = 0. When the wire is stationary, there is no induced emf. As velocity **v,** magnetic induction **B** and length of the wire ℓ are increased, more induced emf is produced.

If the angle between **v** and **B** is θ, the motional emf becomes

$$\varepsilon \quad = \quad -q\, v\, B\, \sin\theta \qquad\qquad 4.4$$

If the resistance of the wire is R, the induced current will be

$$\mathbf{I_{induced}} \quad = \quad \frac{\varepsilon}{R}$$

$$= \quad \frac{-q\, v\, B\, \sin\theta}{R} \qquad\qquad 4.5$$

The unit for motional emf is volts (V).

   Note please: The actual current in wires is due to flow of electrons. The above discussion was based on positive charge carriers.

Example: Determine the induced emf between the wing tips 50 m across of an airplane flying across the earth's magnetic field of 3×10^{-15} T with a velocity of 500 m/s.

$$B_{earth} \quad = \quad 3 \times 10^{-15}\ T$$

$$v \quad = \quad 500\ m/s$$

$$\ell \quad = \quad 50\ m$$

Applying equation 4.3

Induced emf

$$\varepsilon \quad = \quad v\, B\, \ell$$

$$= \quad (500\ m/s)(3 \times 10^{-15}\ T)(50\ m)$$

$$= \quad 75 \times 10^{-12}$$

$$= \quad 7.5 \times 10^{-11}\ V$$

Example: A conductor of length 0.25 m moves with a speed of 5 m/s in a magnetic field of 2500 G. Find the induced voltage in the moving conductor. Also determine the energy delivered to circuit in 2 s if its resistance is 0.75 Ω

$$\ell \quad = \quad 0.25\ m$$

$$v \quad = \quad 5\ m/s$$

$$B \quad = \quad 2500 = 0.25\ T$$

$$t \quad = \quad 2\ s$$

$$R \quad = \quad 0.75\ \Omega$$

$$Emf\ \varepsilon \quad = \quad ?$$

$$Energy \quad = \quad ?$$

Example: (contd.)

Using equation 4.3 $\qquad \varepsilon \quad = \quad -v\,B\,\ell$

$\qquad\qquad\qquad\qquad\qquad = \quad -(5 \text{ m/s}) (0.25 \text{ T}) (0.25 \text{ m})$

$\qquad\qquad\qquad\qquad\qquad = \quad 0.3125 \text{ V (neglecting } - \text{ sign)}$

As resistance R = 0.75 Ω

$\qquad\qquad\qquad I \qquad = \quad \dfrac{V}{R}$

$\qquad\qquad\qquad\qquad\quad = \quad \dfrac{0.3125 \text{ V}}{0.75 \ \Omega} \quad = \quad 0.4166 \text{ A}$

$\qquad\qquad\text{Energy} \quad = \quad (\text{Power}) (\text{time})$

$\qquad\qquad\qquad\qquad\quad = \quad (V\,I)\,(t)$

$\qquad\qquad\qquad\qquad\quad = \quad (0.3125 \text{ V}) (0.4166 \text{ A}) (2 \text{ s})$

$\qquad\qquad\qquad\qquad\quad = \quad 0.26 \text{ J}$

4.4 Faraday's Law and Induced emf

Electromagnetic induction can be easily explained by making use of the notion of **magnetic flux Φ.**

As seen in the previous sections, the electromotive force, ε, was the result of a changing magnetic field cutting through a coil, and the faster a field changed, the greater the resulting emf. Faraday formulated his law of electromagnetic induction using the concept of a changing magnetic flux. The magnetic flux was defined in chapter 3.

Faraday's Law of Electromagnetic Induction states that the emf induced in a loop is proportional to the rate of change of magnetic flux.

Mathematically it is

$$\textbf{emf } \varepsilon \quad \alpha \quad \frac{\Delta\Phi_m}{\Delta t} \qquad\qquad\qquad 4.6$$

where $\Delta\Phi_m$ is the change in magnetic flux during the time interval Δt. The flux Φ_m ($\Delta\Phi_m = \Delta$ **B.A**) would change if B changes, loop area A changes or both the field B and area A change. The values of both B and A are irrelevant. It is the change in flux that is important.

If there are N loops the emf is enhanced N times:

$$\text{emf } \varepsilon \quad = \quad -N \frac{\Delta\Phi_m}{\Delta t} \qquad\qquad\qquad 4.7$$

Faraday's Law can be determined using motional emf ε. When conductor AB is moved from position x_1 to position x_2 (Figure 4.7), the area change ΔA during this motion is

$$\Delta A \quad = \quad (x_2 - x_1)\,\ell \qquad\qquad \text{Area} = L \times B$$

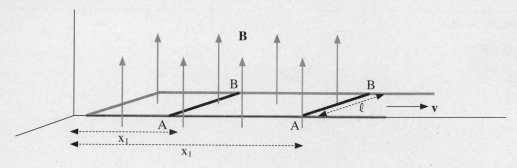

Fig. 4.7: Moving conductor sweeps an area $\Delta A = (x_2 - x_1)\,\ell$ in time Δt.

Induced emf $\qquad\qquad\qquad \varepsilon \quad = \quad v\,B\,\ell$

But $\qquad\qquad\qquad\qquad v \quad = \quad \dfrac{\text{distance}}{\text{time}}$

$\qquad\qquad\qquad\qquad\qquad\quad = \quad \dfrac{x_2 - x_1}{\Delta t}$

and so $\qquad\qquad\qquad\quad \varepsilon \quad = \quad \dfrac{(x_2 - x_1)\,B\,\ell}{\Delta t}$

$\qquad\qquad\qquad\qquad\qquad\quad = \quad \dfrac{(\Delta A)\,(B)}{\Delta t}$

$\qquad\qquad\qquad\qquad\qquad\quad = \quad \dfrac{\Delta (BA)}{\Delta t}$

$\qquad\qquad\qquad\qquad\qquad\quad = \quad \dfrac{\Delta \Phi_m}{\Delta t}$

The above equation is correctly written with a negative (–) sign

$$\varepsilon \quad = \quad -N\,\dfrac{\Delta \Phi_m}{\Delta t}$$

The negative sign indicates that the induced emf and induced current oppose the changing flux (discussed in next section). In case of the motional emf, the induced current retards the motion of the conductor AB.

Example: Find the magnitude of the induced emf when the field through a rotating coil changes from 0 to 0.75 T in 1.2 s. The coil has 25 turns and it is wrapped on a square frame of side 1.5 cm.

$\qquad\qquad N \quad = \quad$ 25 turns

$\qquad\qquad B \quad = \quad$ 0 to 0.75 T

$\qquad\qquad \Delta t \quad = \quad$ 1.2 s

$\qquad\qquad \ell \quad = \quad$ 1.5 cm $\; = \;$ 0.015 m

$\qquad\qquad$ emf $\; = \quad$?

Example: (contd.)

Using equation   emf ε = $-N\dfrac{\Delta\Phi_m}{\Delta t}$

                        = $-N\dfrac{\Delta\,(BA)}{\Delta t}$

But   A = ℓ^2

                        = $(0.015\ \text{m})^2$

                        = $0.000225\ \text{m}^2$

                        = $2.25 \times 10^{-4}\ \text{m}^2$

Inserting values of N, B, A and Δt gives

             emf = $-NA(\Delta B)/\Delta t$

             emf ε = $\dfrac{-(25)\,(2.25\times 10^{-4}\,\text{m}^2)\,(0.75\ \text{T} - 0\ \text{T})}{1.2\ \text{s}}$

                  = $35.16 \times 10^{-4}\ \text{V}$

                  = $3.516 \times 10^{-3}\ \text{V}$

                  = $3.516\ \text{mV}$

4.5 Lenz's Law and Direction of Induced emf

The physicist Heinrich Lenz found that the induced magnetic field (due to the induced current) increased or decreased the field B in such a way as to prevent the change in the magnetic flux. Let us consider the following situation (Figure 4.8).

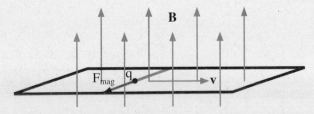

Fig. 4.8: Induced current is out of the plane of the paper.

The field **B** is upward, conductor AB is moved to the right with velocity **v**. The charge q experiences a force $\mathbf{F}_{\text{magnetic}} = q\,(\mathbf{v} \times \mathbf{B})$ out of the plane of the paper and an induced current I flows as a result in the same direction as $\mathbf{F}_{\text{magnetic}}$. This current is directed out of the plane of the paper. The field associated with this current is in concentric circles around it according to the right hand rule (Curl the fingers of the right hand around the conductor AB so that the thumb points in the direction of the current. The fingers curl in the direction of the magnetic field). Thus, there are now two magnetic fields, one **B** and the other due to the current. These fields interact with each other as it is shown in the Figures 4.9a and b.

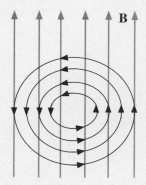

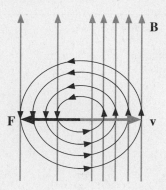

Fig. 4.9a: Field due to induced current (concentric circles) given by right hand rule.

Fig. 4.9b: Interaction of the two magnetic fields: Field cancels on left of conductor; increases on right resulting in a force F (black arrow) towards right opposing **v** (grey arrow).

The field on the left of the conductor cancels out and is reduced while that on the right is enhanced. A force **F** comes into play. This force opposes the change in the flux. It tries to slow down the velocity **v** of the conductor by acting in the opposite direction (Figure 4.9b).

This is known as **Lenz's Law which states that the direction of the induced current is such that its own magnetic field opposes the changes in flux that are inducing it.**

The **microphone is an example of an important application** of electromagnetic induction.

A coil C is attached to the diaphragm D of the microphone. A fixed magnet is placed near the coil so that the coil is in the magnetic field of the magnet (Figure 4.10).

As the sound waves strike the diaphragm D, it vibrates and with it the coil C also moves back and forth thereby changing the magnetic field B across itself (coil C). An induced current is produced in the coil which changes direction as the coil vibrates with the diaphragm.

Thus, sound energy is converted into electric energy (induced current). These electrical signals are sent to amplifiers and then to speakers.

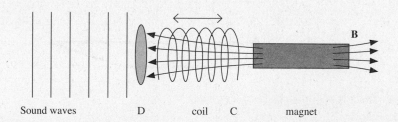

Sound waves   D   coil   C   magnet

Fig. 4.10: A schematic diagram of the microphone.

4.6 Mutual Inductance

As the name suggests, mutual inductance relates two circuits. An emf is produced in the **secondary coil s** due to a changing flux in another coil called the **primary coil p**. A primary coil in which a current

I_p flows produces a magnetic field as shown in the Figure 4.11a. This magnetic field is linked to the secondary coil in such a way so that the field cuts through it.

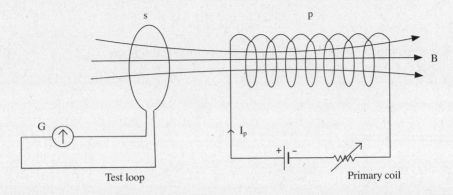

Fig. 4.11a: I_p remains constant __ No induced current.

We already know that the magnetic field produced in a solenoid coil is proportional to the current flowing through it:

$$B \; \alpha \; I$$

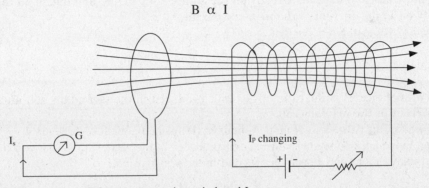

Fig. 4.11b: I_p is increased __ Galvanometer registers induced I_{in}.

Although the secondary (test) coil is placed in the field, no induced current is observed. Can you guess why? As the resistance in the primary circuit is decreased, the current I_p of the primary coil as well as the magnetic field B increases. The magnetic flux through the secondary coil also changes. An emf V_s is induced in the secondary coil and the galvanometer registers induced current I_s (Figure 4.11b). Applying Faraday's Law of electromagnetic induction (equation 4.7):

$$\text{emf}_s \quad \varepsilon_s \quad \alpha \quad \frac{-\Delta\Phi_p}{\Delta t}$$

The rate of change of flux in the primary is proportional to the rate of change of current in the primary. ΔI_p is the current change in the primary coil, thus

$$\text{emf}_s \quad \varepsilon_s \quad \alpha \quad \frac{-\Delta\Phi_p}{\Delta t} \quad \alpha \quad \frac{-\Delta I_p}{\Delta t} \qquad\qquad 4.8$$

and so

$$\varepsilon_s \quad \alpha \quad \frac{-\Delta I_p}{\Delta t}$$

$$\varepsilon_s \quad = \quad -M \frac{\Delta I_p}{\Delta t} \qquad\qquad 4.9$$

M is the constant of proportionality and is called the **mutual inductance**.

Let the number of turns in the secondary be N_s. Applying Faraday's Law

$$\varepsilon_s \quad = \quad -N_s \frac{\Delta \Phi_s}{\Delta t} \qquad\qquad 4.10$$

Equating equations 4.9 and 4.10

$$-N_s \frac{\Delta \Phi_s}{\Delta t} \quad = \quad -M \frac{\Delta I_p}{\Delta t}$$

$$N_s \Phi_s \quad = \quad M I_p$$

$$M \quad = \quad \frac{N_s \Phi_s}{I_p} \qquad\qquad 4.11$$

The mutual inductance M depends directly on the number of turns N_s and flux Φ_s in the secondary and inversely on I_p the current of the primary circuit.

From equation 4.9, it can be seen that

$$M \quad = \quad \frac{-\text{emf}_s}{\Delta I_s / \Delta t} \qquad\qquad 4.12$$

The mutual inductance M is the ratio of the emf_s produced in the secondary coil and the rate of change of current in the primary coil.

It is measured in the SI system of units in **henrys (H)** named after Joseph Henry.

The mutual inductance is one henry if the emf produced in the secondary is one volt due to rate of change of current equal to one ampere per second in the primary coil.

$$1H \quad = \quad \frac{1V}{1A/s} \qquad\qquad 4.13$$

In calculations **mH (10^{-3}H) and μH (10^{-6} H)** are commonly used.

The mutual inductance M depends on the geometry of coils and the core material used.

Example: A loop of wire is placed near a solenoid. The flux through the loop is 2.5×10^{-5} Wb when the current in the solenoid is 0.5 A. The current in the solenoid is changing at the rate of 5 A/s and the induced current in the loop is 0.35 mA. What is the resistance of the loop?

$$\Phi_s \quad = \quad 2.5 \times 10^{-5} \text{ Wb}$$

$$I_p \quad = \quad 0.5 \text{ A}$$

$$\Delta I_p / \Delta t \quad = \quad 5 \text{ A/s}$$

$$I_s \quad = \quad 0.35 \text{ m A} \quad = \quad 0.35 \times 10^{-3} \text{ A}$$

Example: (contd.)

$$R = ?$$

$$R = \frac{V}{I} = \frac{emf_{induced}}{I_{induced}}$$

$$\varepsilon_s = -M\,(\Delta I_p/\Delta t) \qquad \text{(equation 4.9)}$$

$$= -M\,(5\text{ A/s})$$

For one loop

$$\Phi_s = M\,I_p \qquad \text{(equation 4.11)}$$

$$M = \frac{\Phi_s}{I_p}$$

$$= \frac{2.5\times 10^{-5}\,\text{Wb}}{0.5\text{ A}}$$

$$= 5 \times 10^{-5}\text{ H}$$

Therefore

$$\varepsilon_s = -(5 \times 10^{-5}\text{ H})\,(5\text{ A/s})$$

$$= -25 \times 10^{-5}\text{ V}$$

and resistance of secondary coil is

$$R = \frac{\varepsilon_s}{I_s}$$

$$= \frac{-25\times 10^{-5}\text{ V}}{0.35\times 10^{-3}\text{ A}}$$

$$= 7.14 \times 10^{-1}\ \Omega \text{ (neglecting – sign)}$$

$$= 0.714\ \Omega$$

4.7 Self Inductance

We have seen in all the above examples of electromagnetic inductance a changing external field produces an induced emf in a coil placed in the field.

In section 4.6, the flux in the primary coil was also changing. Therefore an emf should be induced in the primary coil as well according to Faraday's Law. Such an effect when the rate of change of flux in the coil induces an emf in the coil itself is called **self inductance**.

Consider the circuit in Figure 4.12. As the current is changed by increasing/decreasing the resistance, the magnetic flux changes in the circuit coil. This results in an induced emf in the coil.

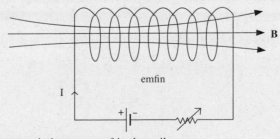

Fig. 4.12: The changing current induces an emf in the coil.

The flux Φ_m is proportional to the magnetic field B and B itself is directly proportional to I for a solenoid $(B = \mu_0 nI)$. For N turns

$$N\Phi_m \quad \alpha \quad B \quad \alpha \quad I$$

$$N\Phi_m \quad \alpha \quad I$$

$$N\Phi_m \quad = \quad LI$$

where L is the constant of proportionality and is called **self inductance**.

$$L \quad = \quad \frac{N\Phi_m}{I} \qquad 4.14$$

Using Faraday's equation

$$emf \; \varepsilon \quad = \quad -N\frac{\Delta\Phi_m}{\Delta t}$$

$$emf \; \varepsilon \quad \alpha \quad -N\frac{\Delta\Phi_m}{\Delta t}$$

From equation 4.14

$$N\Phi_m \quad = \quad LI$$

$$emf \; \varepsilon \quad \alpha \quad \frac{-\Delta(LI)}{\Delta t}$$

$$\mathbf{emf} \; \boldsymbol{\varepsilon} \quad = \quad -L\frac{\Delta I}{\Delta t} \qquad \qquad L = constant \qquad 4.15$$

The self inductance L is

$$L \quad = \quad \frac{\varepsilon}{\Delta I/\Delta t} \qquad 4.16$$

The self inductance is the ratio of the emf induced in a coil due to the rate of change of current in the coil.

The unit of self inductance is henry (H) in SI system.

The self inductance also depends on the geometry of the coil and the core material. The self induction can be increased by placing an iron core in the coil. Inductor coils are used as chokes in circuits. Because of the induced emf they produce, they oppose (– sign) both increasing and decreasing currents in circuits.

Suppose we increase the current in a circuit containing an inductor coil. The induced current will oppose this increase. It will flow in the opposite direction (Figure 4.13). On the other hand, if we try to decrease the current in the circuit, the back emf will produce a current in the same direction. The induced emf is also termed **back emf** because it opposes the change that produces it.

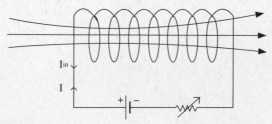

Fig. 4.13: Circuit showing induced current I_{in} when current I is being increased.

4.7 Energy stored in an Inductor

The inductor coil can store energy. As the voltage from the source is increased, current increases and a back emf $\varepsilon = -L\,(\Delta I/\Delta t)$ opposes this increase (Section 4.6). To keep on increasing the current in the circuit, the generator has to do more work. The work (ΔW) done by the generator to move a charge Δq around the loop is given by

$$\Delta W \quad = \quad \Delta q\ \varepsilon \qquad\qquad\qquad \text{as } W = q\ V$$

$$= \quad \Delta q\ \left[L\frac{\Delta I}{\Delta t}\right]$$

$$= \quad L\left[\frac{\Delta q}{\Delta t}\right]\Delta I$$

$$= \quad L\ I\ (\Delta I) \qquad\qquad\qquad\qquad\qquad 4.17a$$

ΔW is the work done by the generator to increase the current by a value ΔI. The total work can be computed by adding all ΔWs as current goes from zero to a maximum I.

As the induced current increases from 0 to a maximum value I, ΔI can be taken as the average current.

The average current is given by $\qquad I_{average} \quad = \quad \dfrac{0+I}{2} \quad = \quad \dfrac{I}{2}$

Total work

$$\mathbf{W} \qquad = \qquad \tfrac{1}{2}\,\mathbf{L\ I^2} \qquad\qquad\qquad\qquad 4.17b$$

By using the calculus and integrating the equation 4.17a, the same result can be obtained:

$$\int\!\Delta W \quad = \quad \int L\ I\ (\Delta I)$$

$$W \quad = \quad \tfrac{1}{2}\,L\ I^2$$

The work done is equal to the energy stored in the inductor. It is analogous to the more familiar equation for energy ($\tfrac{1}{2}\,mv^2$). Where is this energy stored? In the capacitor, energy was stored in the electric field and was directly proportional to the square of the electric field intensity (E^2). In the inductor, energy is stored in its magnetic field and can be derived from:

$$L \quad = \quad \frac{N\Phi}{I}$$

$$= \quad \frac{N\,(BA)}{I} \qquad\qquad \text{as } \Phi = BA$$

$$= \quad \frac{N\,(\mu_o nI)\,A}{I} \qquad\qquad B = \mu_o n\,I$$

As the number of turns per unit length is $n = N/\ell$

$$L \quad = \quad (n\ell)\, \mu_o\, n\, A$$

$$\mathbf{L} \quad = \quad \boldsymbol{\mu_o}\, \mathbf{n^2}\, \mathbf{A}\, \ell \qquad\qquad 4.18$$

$$\text{Energy } U \quad = \quad \tfrac{1}{2}\, L\, I^2$$

$$= \quad \tfrac{1}{2}\, \mu_o\, n^2\, A\ell\, I^2$$

So $\qquad\qquad$ Energy $U \quad = \quad \tfrac{1}{2}\, (\mu_o\, n^2\, A\ell) \left[\dfrac{B}{\mu_o n}\right]^2 \qquad\qquad I = \dfrac{B}{\mu_o n}$

$$= \quad \tfrac{1}{2}\, \dfrac{A\ell\, B^2}{o}$$

Energy density $\qquad\qquad = \quad$ Energy/volume

$$= \quad \dfrac{1}{2\mu_o}\, B^2 \qquad\qquad 4.19$$

Energy density $\qquad\qquad \alpha \quad \mathbf{B^{\,2}} \qquad\qquad 4.20$

The energy stored is proportional to the square of the magnetic induction B.

Example: Calculate the inductance of solenoid of length 40 cms and having 500 turns. Its radius of cross-section is 1.25 cm.

$$\mu_o \quad = \quad 2 \times 10^{-7}\ \text{T.m/A}$$

$$\ell \quad = \quad 40\ \text{cms } = 0.4\ \text{m}$$

$$r \quad = \quad 1.25\ \text{cm} = 1.25 \times 10^{-2}\ \text{m}$$

$$N \quad = \quad 500\ \text{turns}$$

$$L \quad = \quad ?$$

$$\text{Using } L \quad = \quad \dfrac{\mu_o N^2\, A}{\ell}$$

$$A \quad = \quad \pi\, r^2$$

$$= \quad 3.14 \times (1.25 \times 10^{-2}\ \text{m})^2$$

$$= \quad 4.906 \times 10^{-4}\ \text{m}^2$$

$$L \quad = \quad \dfrac{\mu_o N^2\, A}{\ell}$$

$$= \quad \dfrac{2 \times 10^{-7}\,\text{T.m/A}\, (500)^2\, (4.906 \times 10^{-4}\,\text{m}^2)}{0.4\ \text{m}}$$

$$= \quad \dfrac{2 \times 25 \times 4.906 \times 10^{-7}}{0.4\ \text{m}}$$

$$= \quad 613.25 \times 10^{-7}\ \text{henrys}$$

$$= \quad 61.325\ \mu\text{H}$$

Test yourself: If the current in the above solenoid is decreasing at 100 A/s, what is the induced emf?

4.8 Electric Generator (Alternating current or AC generator)

Faraday's Law of electromagnetic induction led to the development of the electric generator and the transformer which play such important roles in our lives today. Practically all the world's electrical energy is produced by the application of this law.

Electrical energy is primarily produced by electric generators. The principle underlying the electric generator is the concept of motional emf.

In an electric generator, a coil, the armature, is rotated in a magnetic field. The coil is attached to a shaft that can be rotated by a turbine. Consider a single rectangular loop (a coil with one turn only) PQRS placed in a uniform magnetic field B (Figure 4.14). It is rotated at an angular velocity ω by the turbine. As the coil rotates, the end termini move over circular rings which slide on fixed carbon brushes connected to the leads of the external circuit.

As the loop is rotated, the flux changes through it and an induced emf is produced. The motional emf ε across arm PQ is

$$\text{emf } \varepsilon_{PQ} \quad = \quad v\, B\, \ell \sin \theta$$

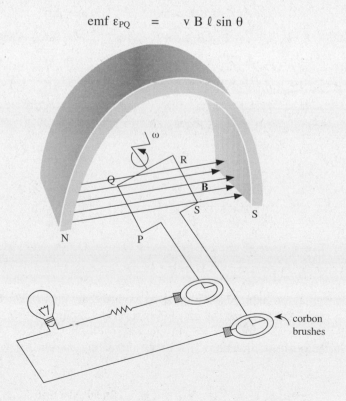

Fig. 4.14: An AC generator.

As arm PQ moves upwards, arm RS moves downwards, an equal but opposite induced emf is produced across PQ and RS.

No contribution to the emf is made by the sides QR and SP. Consider QR. As the coil is rotated on an axis through the centre of QR, half the arm goes upwards and the other half downwards. Any emf produced in one half QR will be equal and opposite to that produced in the other half. It is so cancelled.

As the conductor is in the form of a rectangular loop, the emfs from sides PQ and RS add up and so the total emf produced is

$$\text{emf}_{\text{total}} \quad = \quad \text{emf } \varepsilon_{PQ} \quad + \quad \text{emf } \varepsilon_{RS}$$

$$= \quad vB\ell\sin\theta \quad + \quad vB\ell\sin\theta$$

$$= \quad 2\, v\, B\, \ell\, \sin\theta$$

$$\text{emf } \varepsilon_{\text{total}} \quad = \quad 2\,(\omega\, r)\, B\ell\, \sin\theta \qquad\qquad \text{as } v = r\,\omega$$

But r = b/2 where b is the breadth of the coil:

$$\text{emf } \varepsilon_{\text{total}} \quad = \quad 2\,\omega\,(b/2)\, B\ell\, \sin\theta$$

$$= \quad 2\pi f\, B\, b\ell\, \sin\theta \qquad\qquad \omega = 2\pi f$$

$$= \quad 2\pi\, f\, B\, A\, \sin\theta \qquad\qquad A = \text{area of coil}$$

$$= b\ell$$

For N turns

$$\text{emf } \varepsilon_{\text{total}} \quad = \quad 2\pi\, f\, BAN\, \sin\theta$$

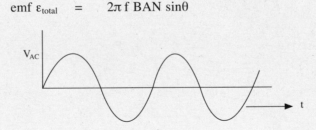

Fig. 4.15: The voltage changes direction every half cycle.

The total emf $\varepsilon_{\text{total}}$ as well as the induced AC current so produced are sinusoidal (Figure 4.15). As the coil moves through a cycle, the force on the arm PQ is reversed as the arm PQ first moves upwards and then downwards. This is also true for RS. In one half-cycle, the induced current flows in one direction, and then in the opposite direction. This is the alternating current AC voltage and will now be written as V_{AC}.

$$\text{emf } \varepsilon_{\text{total}} \quad = \quad \mathbf{V_{AC}} \quad = \quad \mathbf{2\pi\, f\, BAN\, \sin\theta} \qquad\qquad 4.21$$

The AC voltage is maximum when $\sin\theta = 1$ and $\theta = 90°$ and equal to

$$\mathbf{V_o} \quad = \quad \mathbf{2\pi\,f\ BAN} \qquad\qquad 4.22$$

Equation 4.22 becomes $\qquad\qquad \mathbf{V_{AC}} \quad = \quad \mathbf{V_o\ \sin\theta} \qquad\qquad 4.23$

When $\theta = 0°$, $\sin\theta = 0 \qquad\qquad \mathbf{V_{AC}} \quad = \quad \mathbf{0} \qquad\qquad 4.24$

The magnitude of the current depends on the angle between **v** and **B**. Thus the current goes from zero to a maximum and back to zero like a sine curve, reverses its direction and again goes to a maximum and then to zero. The cycle is repeated (Figure 4.16).

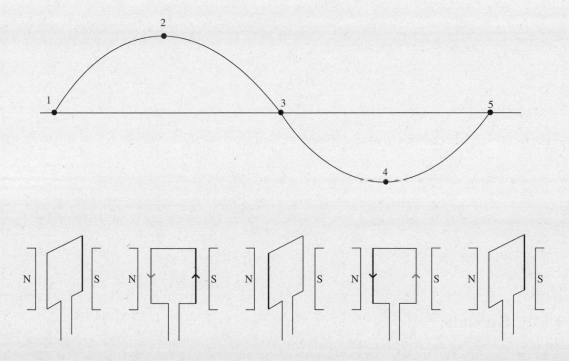

Fig. 4.16: V_{AC} at different positions of the cycle.

The frequency of the AC generator in Pakistan is 50 c/s or 50 Hz.

In hydroelectric power plants, the generator shaft is rotated by turbines driven by the potential energy of falling water. In coal-powered plants, a steam engine drives the shaft.

Example: An AC generator coil has 20 turns and area of each coil is 0.1 m². Its resistance is 20 Ω. It is placed in a field of 0.5 T and the frequency of rotation is 50 Hz. What is the maximum induced emf, maximum induced current and the induced emf in the coil?

$$N \quad = \quad 20 \text{ turns}$$
$$A \quad = \quad 0.1 \text{ m}^2$$
$$B \quad = \quad 0.5 \text{ T}$$
$$R \quad = \quad 20 \text{ Ω}$$
$$f \quad = \quad 50 \text{ Hz}$$

The maximum induced emf from equation 4.24

$$V_o \quad = \quad N\,A\,B\,2\,\pi\,f$$
$$= \quad (20)\,(0.1 \text{ m}^2)\,(0.5 \text{ T})(2\,\pi)\,(50)$$
$$= \quad 314 \text{ V}$$

Maximum induced current

$$I_{max} \quad = \quad \frac{V_o}{R}$$
$$= \quad \frac{314 \text{ V}}{20 \text{ Ω}}$$
$$= \quad 15.7 \text{ A}$$
$$\text{Induced emf} \quad = \quad V_o \sin(2\,\pi\,f\,t)$$
$$= \quad 314.0 \sin(2\,\pi\,f\,t)$$
$$= \quad 314 \sin(314\,t)$$

4.9 DC Generator

In the AC generator, the induced emf and induced current are sinusoidal functions of time as is observed from the equation for the induced emf

$$V_{AC} \quad = \quad 2\pi\,f\,BAN \sin\theta$$

The current and voltage change direction every half-cycle (Figure 4.15).

To get a current that is in one direction, a commutator (split ring) is used instead of the circular rings used in AC generator connections to the external circuit (Figure 4.17).

As the coil rotates and current changes direction every half-cycle, the leads from the generator to external wire first connect to one half and then to the other half of the split ring. And so the current flows in one direction in the external circuit. The output waveform is in one direction as shown in Figure 4.18.

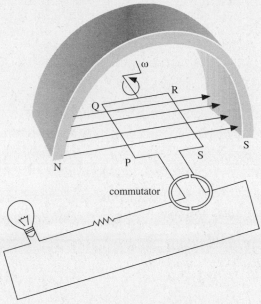

Fig. 4.17: A DC generator — external circuit is connected through split ring.

To remove the AC element (fluctuation) from the DC output, a number of coils are wired in series. These coils are placed at different positions. For instance, if two are placed perpendicular to one another (Figure 4.19a) when one gives zero current, the other will give maximum current. The output currents add up and the fluctuation is reduced.

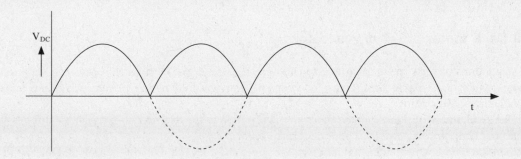

Fig. 4.18: DC generator output.

As seen in Figure 4.19b, the DC output from the two perpendicular coils has reduced the fluctuation. As more of these coils are placed at different angles, fluctuation in current is further reduced.

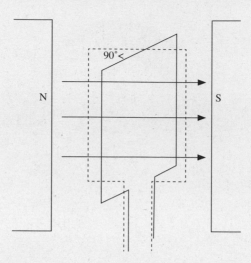

Fig. 4.19a: Two coils placed perpendicular to one another.

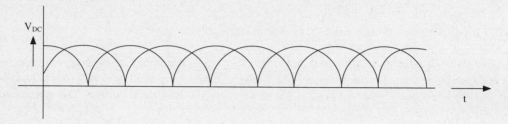

Fig. 4.19b: Output of a DC generator when two coils are placed perpendicular to each other.

4.10 Back motor effect in generators

The generator converts mechanical energy into electrical energy. Mechanical energy is supplied by turbines which are rotated themselves by steam in coal powered power plants; or falling water in hydroelectric power plants.

The electrical energy produced in the generator is supplied to the external circuits. These circuits have resistances and consume energy. They are termed **load.**

When the load is not drawing any current, the generator does not feel any burden and energy losses are minimized. They are only due to overcoming friction in turbines and generators.

When current is drawn by the load, the generator too has a current in its coils. When a changing current flows in a conductor placed in a magnetic field, an opposing force (refer section 4.5) is produced in the arms AB and CD. This results in an opposing torque which increases with current. Because of this torque which tends to make the coil move in the opposite direction to the shaft angular motion, the turbine then would need more power to turn the shaft and coil. Thus, more fuel would have to be expended to give the turbines more energy to rotate the coil at the same angular speed.

The opposing torque is termed the back motor effect in generators.

This is a good example of the law of conservation of energy. The energy (used up) by the external circuit (load) comes from the basic energy source (coal, gas, water, etc) that drives the turbines.

4.11 DC Motors

The DC motor converts electric energy to mechanical energy.

The principle of the DC motor was given in Chapter 3. When a current-carrying conductor is placed in a magnetic field, it experiences a torque and is able to rotate.

Consider a rectangular loop of length ℓ and breadth 2r placed in a uniform magnetic field B. Externally the coil is connected to a DC voltage source V. A current I flows through the circuit. As the current flows in the coil, a downward force ($\mathbf{F} = I\ell \times \mathbf{B}$) is produced on the arm 1 and an upward force on arm 2 resulting in a torque that makes the coil rotate. Sides 2 and 4 do not contribute to the torque.

The problem was that as the arms 1 and 2 change direction, the torque direction would reverse and the coil would not be able to rotate and function as a motor.

To overcome this problem so that the arms moving on the left always have current flowing into the plane of the paper and those on the right should have current moving out of the plane of the paper, a **commutator or split ring** is utilized. As soon as the coil comes to the position at which its direction changes, the leads of the coil move from one half of the split ring to the other half. In this way, the coil keeps on rotating (Figure 4.20).

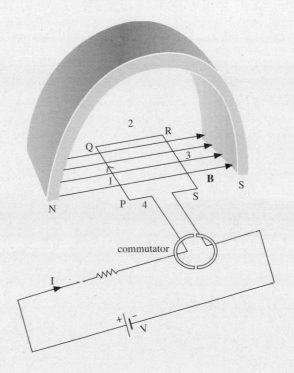

Fig. 4.20: DC motor.

4.12 Back emf in Motors

As discussed in the previous section, when current passes through a commutator ring to a coil placed in a magnetic field, the coil can rotate. When the coil rotates, flux changes through it and an emf ε which is opposite in polarity to the battery emf V is produced in the coil. This emf is called **back emf**. The net emf as well as the current flowing through the coil is reduced:

$$\text{Net emf} = \text{emf}_{\text{battery}} - \text{emf}_{\text{back}}$$
$$= V - \varepsilon$$

If the resistance of the coil is R then the current in the coil is equal to

$$I = \frac{\text{Net emf}}{R}$$

$$I = \frac{V - \varepsilon}{R} \qquad\qquad 4.25$$

At the time the motor is started, the current in the coil is higher and equal to

$$I_{max} = \frac{V}{R} \qquad\qquad 4.26$$

As the coil rotates faster and faster the induced back emf keeps on increasing and so the current in the coil is reduced. The current has to be enough to rotate the coil and the load attached to it. If the load is too heavy, the motor will slow down. As it slows down, the back emf reduces and current in the coil rises.

Too much overloading may cause very high currents which may melt the coil wire and destroy the motor.

The back emf can be shown by connecting a motor in series with a light bulb (Figure 4.21). When the switch is closed, the lamp is bright initially. There is zero back emf at the start. The lamp soon dims as the motor speeds up and a back emf develops. If a heavy load is placed on the motor, it slows down and the lamp again becomes bright.

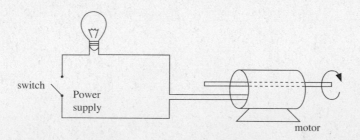

Fig. 4.21: A motor is connected to a power supply and a bulb.

4.13 Transformers

Transformers are important devices based on electromagnetic induction that have helped save millions of rupees worth of electrical energy. They are used in daily life to decrease or increase voltage. Electricity that is consumed in Lahore for instance is produced in Mangla. It is transported through wires to Lahore.

We know already that power dissipated in a resistor R is equal to I^2R. To reduce power losses the current I or/and the resistance R must be reduced.

Reducing R: Resistance depends on resistivity ρ, length of the conductor ℓ and cross-sectional area A of the conductor

$$R \quad = \quad \rho\,\ell/A$$

which suggests that to decrease the resistance R, length ℓ and resistivity ρ should be decreased and area A should be increased.

To decrease the resistivity, a material of low resistivity like copper should be used for wires. Wires are mainly of copper. The length ℓ, in this case, is the distance between Mangla and Lahore which can only be reduced to the shortest distance between the two places. The area A of the wire must be large to decrease resistance. Copper is an expensive metal so increasing the area is not economically viable. Another reason for not increasing area A is that by increasing the area, the wires become not only more expensive but very heavy as well. Thus, there is a limitation to reducing R.

In reducing the other factor I is where transformers play an important role.

Transformers are constructed using an iron core and two coils, a primary coil that has an AC source and a secondary coil that gives the output across a load resistance. The wires of the coils are insulated and wrapped over each other ensuring that the change in the magnetic flux in one is communicated totally into the other (Figure 4.22).

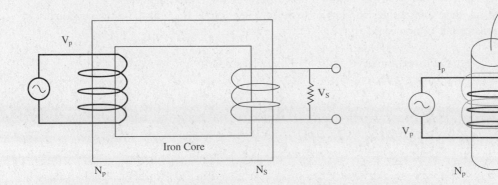

Fig. 4.22: Two schematic diagrams of transformers.

Using Faraday's Law of electromagnetic induction (equation 4.6), the emf in the primary coil is equal to

$$\text{emf}_p \quad = \quad V_p \quad = \quad -N_p\,\frac{\Delta\Phi}{\Delta t} \qquad\qquad 4.27$$

And the emf in the secondary is given by

$$\text{emf}_s \quad = \quad V_s \quad = \quad -N_s \frac{\Delta\Phi}{\Delta t} \qquad\qquad 4.28$$

As the rate of change of flux is the same through both the primary and secondary coils, equations 4.27 and 4.28 can be equated to

$$\frac{V_p}{N_p} \quad = \quad \frac{V_s}{N_s} \qquad\qquad 4.29$$

Equation 4.29 shows that the output from the secondary coil of the transformer is

$$V_s \quad = \quad \frac{N_s}{N_p} V_p \qquad\qquad 4.30$$

The output voltage from the secondary coil of a transformer is equal to the product of the primary emf V_p and the ratio of the turns of the secondary N_s and that of the primary N_p.

It can be seen that if

$$N_s \quad > \quad N_p$$

$$V_s \quad > \quad V_p$$

When a transformer gives a higher output it is called a **Step-up Transformer.**

And if

$$N_s \quad < \quad N_p$$

$$V_s \quad < \quad V_p$$

The output voltage is less than the input voltage. It is a **Step-down Transformer.**

Assuming it is an ideal transformer and power losses are negligible, then all the power produced in the primary is transferred to the secondary:

$$\text{Power primary} \quad = \quad \text{Power secondary}$$

$$P_p \quad = \quad P_s$$

$$\mathbf{V_p\, I_p} \quad = \quad \mathbf{V_s I_s} \qquad\qquad 4.31$$

or $\qquad\qquad\qquad \dfrac{V_s}{V_p} \quad = \quad \dfrac{I_p}{I_s} \qquad\qquad 4.32$

Equation 4.32 shows that current and voltage are inversely related in transformers. When voltage is increased, current is reduced and vice versa. Ohm's Law is not valid in transformers.

Our second consideration for power transmission was current reduction. In Mangla, if we use a step-up transformer we can drastically reduce the current.

We have step-down transformers near our homes that can bring the voltage down to 220~240 V and currents up to 3.5 A. You must have realized now why transformers have the **DANGER** sign.

We use transformer-like devices in the home in charging mobile phones, in contact lens cleaners, in cassette players and radios. In all these devices, the voltages needed are small [9V – 24V] and so these **adapters** are step-down transformers.

Example: A step-up transformer having 400 turns in the primary is connected to a 220V power supply. The current in the primary 5A and the number of turns in the secondary coil are 2000. What is the current in the secondary?

$$V_p \quad = \quad 220 \text{ V}$$

$$I_p \quad = \quad 5 \text{ A}$$

$$N_p \quad = \quad 400 \text{ turns}$$

$$N_s \quad = \quad 2000 \text{ turns}$$

$$I_s \quad = \quad ?$$

The current in the secondary I_s is given by

$$I_s \quad = \quad I_p \, (V_p/V_s)$$

I_p and V_p are given. Using equation 4.30 to find V_s

$$V_s \quad = \quad V_p \, (N_s/N_p)$$

$$= \quad 220 \, (2000/400)$$

$$= \quad 1100 \text{ V}$$

and so $\qquad\qquad I_s \quad = \quad (5A) \, (220V/1100V)$

$$= \quad 1 \text{ A}$$

Test yourself: A lady buys an expensive hair dryer in the USA [V =110V]. She wants to use it in Pakistan and so buys a transformer with 100 turns in the secondary coil. How many turns should there be in the primary coil?

4.13.1 Transformer Efficiency

In ideal transformers, all primary power is transferred to the secondary circuit. In practice, however, the efficiency of the transformer although high is about 95%. Efficiency is defined as the ratio of the power in the secondary to the power in the primary.

$$\text{Efficiency} \quad = \quad \frac{\text{Power}_{secondary}}{\text{Power}_{primary}} \quad \times \quad 100\%$$

$$= \quad \frac{P_s}{P_p} \quad \times \quad 100\%$$

4.13.2 Transformer Losses

Although transformer efficiency is high, the transfer of power from the primary circuit to the secondary does not occur without some power losses. The losses in transfer of power in transformers are due to:

1. **Eddy Current Losses**

   The transformer coils are wound on iron core. As the flux changes in the coil it also changes in the iron mass. Induced currents are produced which circulate in the iron masses. These currents produce heat (I^2R), power is dissipated and they do no useful work. They are called **Eddy currents.**

   These eddy currents losses can be reduced by laminating the armature frames and cores. Thin insulated metal sheets are used to make transformer cores. The metal sheets are placed so that the eddy currents are restricted to the width of the sheets. The induced currents and related heat losses are reduced.

2. **Copper Losses**

   Transformer coil wires are made up of copper. If the wire resistance is R, then as current I flows through them, heat is dissipated as I^2R. These are unavoidable losses.

3. **Hysteresis Losses**

   Some power is lost because the transformer core gets magnetized and demagnetized again and again due to the AC current in the coils wound around the core.

Summary

The magnetic flux Φ through a closed loop is defined as

$$\Phi_B = BA \cos \theta$$

where B is the strength of the uniform magnetic field, A is the cross-sectional area of the loop, and θ is the angle between B and the direction perpendicular to the plane of the loop.
Induced emf characteristics are:

1. Moving a loop across a magnetic field that cuts across the loop produces an induced emf and induced current in the loop.
2. When the direction of motion is reversed, the current flows in the opposite direction.
3. The faster the loop is moved the greater the current.
4. The current is more when field is increased.

If a conducting bar of length ℓ moves through a magnetic field with a speed v, so that **B** is perpendicular to the bar, the emf induced in the bar, often called a **motional emf,** is

$$\varepsilon = -vB\ell$$

Faraday's Law of Induction states that the instantaneous emf induced in a circuit equals the rate of change of magnetic flux through the circuit:

$$\varepsilon = N \frac{\Delta \Phi_B}{\Delta t}$$

where N is the number of loops in the circuit.

Lenz's Law states that the polarity of the induced emf is such that it produces a current whose magnetic field opposes the change in magnetic flux through a circuit.

The mutual inductance M depends directly on the number of turns N_s and flux Φ_s in the secondary and inversely on I_p the current of the primary circuit.

$$M = \frac{N_s \Phi_s}{I_p}$$

The mutual inductance M can also be defined as the ratio of the emf_s produced in the secondary coil and the rate of change of current in the primary coil.

$$M = \frac{-\mathrm{emf}_s}{\Delta I_p / \Delta t}$$

It is measured in the SI system of units in **henrys (H)** named after Joseph Henry.

Emf ε is also induced in the coil itself when current changes in the coil:

$$\varepsilon = L \frac{\Delta I}{\Delta t}$$

where L is the **self inductance** of the coil. Self inductance is also measured in henries (H):

$$1H = 1 \frac{Vs}{A}$$

The **inductance** of a coil can be found from the expression

$$L = \frac{N \Phi_B}{I}$$

where N is the number of turns on the coil, I is the current in the coil, and Φ_m is the magnetic flux through the coil produced by that current.

The energy stored in the magnetic field of an inductor carrying current I is

$$\text{Energy} = \frac{1}{2} LI^2$$

The energy stored in an inductor is proportional to the square of the magnetic induction B:

$$\text{Energy density} \quad \alpha \quad B^2$$

Alternating current (AC) voltage V_{AC} is produced in an AC generator. It is equal to

$$V_{AC} \;=\; 2\pi f \, BAN \sin\theta$$

Commutator rings are used to convert AC into DC in DC generators. To remove AC fluctuations, two coils are placed perpendicular to each other.

The opposing torque is termed the **back motor effect** in generators.

DC motors are an application of torque produced by current-carrying coils placed in a magnetic field. Electrical energy is converted into mechanical energy.

When the coil rotates, the flux changes through it and an emf ε which is opposite in polarity to the battery emf V is produced in the coil. This emf is called **back emf**.

Transformers are step-up or step-down voltages. Transformer equations are:

$$\frac{V_s}{V_p} \;=\; \frac{N_s}{N_p}$$

and

$$\frac{V_s}{V_p} \;=\; \frac{I_p}{I_s}$$

The **losses in transfer of power in transformers** are due to:

1. Eddy Current Losses
2. Copper Losses
3. Hysteresis Losses

Questions

1. When is the induced emf maximum?
2. How can we place two current-carrying wires to reduce magnetic fields between the wires?
3. Why is a commutator used in an electric motor?
4. What is the relation between the direction of an induced current and the motion which produced it?

5. A loop is placed in a uniform magnetic field. For what orientations of the loop is the magnetic flux maximum? For what orientation is the flux zero?
6. Where is energy stored in inductors?
7. A loop of copper wire is rotated in a magnetic field about an axis along a diameter. Why does the loop resist motion?
8. Why is it easy to turn the shaft of a generator when it is not connected to an outside circuit, but much more difficult when an external circuit is connected?
9. Why is the induced emf that appears in a conductor called a back emf?
10. What is the value of the back emf when the motor is started?
11. If the current in an inductor is doubled by what factor would the energy change?
12. Show that $L = \dfrac{N\Phi_m}{I}$ and $L = -\dfrac{\varepsilon}{\Delta I/\Delta t}$ have the same units.
13. Can transformers be used to increase/decrease dc voltage?

Problems

1. The frequency of AC supply in Pakistan is
   a) 50 Hz   b) 60 Hz
   c) 100 Hz   d) none of these
2. The SI unit of mutual inductance is
   a) Tesla   b) Henry
   c) Weber   d) Farad
3. A circular coil of area 4 m^2 is placed in a magnetic field B of 0.75 T. What would be the magnetic flux for $\theta = 45°$?
4. A p.d. of 4 V is found between the ends of a 2 m wire moving perpendicular to the field. The speed of the wire is 20 m/s. What is the magnitude of the field?
5. A car is moving with a speed of 50 m/s on a road where the earth's magnetic field is 8×10^{-5} T downwards. What is the p.d. between the ends of its axles which are 1.5 m long?
6. A circular coil of radius 15 cm is placed in an external magnetic field of 0.25 T so that

the plane of the coil is perpendicular to the field. The coil is pulled out of the field in 0.3 s. Find the average induced emf during this interval.

7. Two coils have a mutual inductance of 4 mH. In the primary coil, the current is changing by 3.6 A in 0.03 s. The circuit with the secondary has a resistance of 1.5 Ω. Find the magnitude of the average current induced in the secondary coil.

8. A coil has 275 turns and a self inductance of 0.015 H. The coil carries a current of 0.02 A. Obtain the magnetic flux through one turn of the coil.

9. The current in a circuit falls from 5 A to 1 A in 0.1s. If an average emf of 2 V is induced in the circuit while this is happening, find the inductance in the circuit.

10. A rectangular wire loop 5 cm × 10 cm in size is perpendicular to a magnetic field of 1×10^{-3} T.
    a) What is the flux through the loop?
    b) If the magnetic field drops to zero in 3 s, what is the potential difference induced between the ends of the loop during that period?

11. In the Figure, R = 10 Ω and ℓ = 1.5 m. A uniform magnetic field of 3 T is directed into the plane of the paper. At what speed should AB move so that a current of 0.67 A is produced in R?

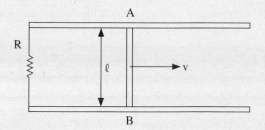

12. How much energy is stored in a 75 mH inductor at an instant when the current is 2.5 A?

13. A solenoid of length 10 cm has 95 turns. Its diameter is 1.5 cm and it has an air core.

How much energy is stored if the current in the solenoid is 0.75 A?

14. An inductor of self inductance 250 μH can allow a maximum current of 1.2×10^6 A to flow through it. What is the energy stored in the inductor when maximum current flows through it and what is the power during 10 ms, the time in which the inductor discharges?

15. A circular loop is placed near a coil. When current in the coil is 550 mA, the flux through the loop is 2.7×10^{-5} Wb. When the current in coil changes at the rate of 6 A/s, the induced current in the circular loop is 0.36 mA.
    a) What is the mutual inductance?
    b) What is the resistance of the circular loop?

16. If a force of 5 N is required to move a conductor through a magnetic field at the rate of 2 m/s while 1 V is being generated, how many amperes are flowing through the conductor? (Use power equation).

17. A man buys an electric shaver in USA where the voltage is at 110 V. He wants to use it in Pakistan. To convert to the necessary voltage, he buys a transformer with 550 turns in the primary coil.
    a) Is it a step-up or step-down transformer?
    b) How many turns are there in the secondary coil?

18. A step-down transformer (turns ratio = 1 : 8) is used with an electric toy train to reduce the voltage from the wall receptacle (220V) to a value needed to operate the train. When the train is running the current in the secondary coil is 3.4 A. What is the current in the primary coil?

19. A coil having 100 turns and area 0.05 m^2 rotates about its axis at 1200 revs/min. Compute the maximum emf if the magnetic field through the coil is 3×10^{-5} T.

20. When rotating at 500 revs/min. a car generator produces 12 V. What potential difference will it produce at 1300 revs/ min?

5 Alternating Current

OBJECTIVES

After studying this chapter, students should be able:
*   to understand that AC current and voltage are sinusoidal and its frequency, time period and phase can be determined from the waveform.
*   to understand why AC is measured in root-mean square values.
*   to understand that AC can flow in capacitors unlike DC.
*   to understand that AC current and voltage lead or lag each other in inductors and capacitors.
*   to understand that resonance occurs in AC circuits and its importance in transmission and receiving sound and picture electromagnetic signals.
*   to understand how LC circuits can be used in metal detectors and detectors.
*   to understand the difference between x-rays, uv rays, light, radio waves etc. in the electromagnetic spectrum.
*   to understand how electromagnetic waves are used in transmission and modulation of sound and picture signals.

5.1 Alternating Current

In the last chapter, we saw that alternating current (AC) is generated by a generator and is **sinusoidal.** The equation for AC emf is

$$\Delta V_{AC} \quad = \quad 2\pi \ f \ BAN \ \sin\omega t \qquad\qquad 5.1$$

It can be seen that **AC voltage varies with time.** The minimum value is zero at $\sin 0°$ and maximum value is equal to $2\pi f \ BAN$ at $\sin 90°$. The angular velocity is ω and it is equal to $2\pi f$, where f is the frequency measured in c/s or hertz (Hz).

The sinusoidal emf given above shows that the AC voltage and the current change polarity twice during each cycle. As shown in section 4.8, the coil is in position 1 at time t = 0. If the time period or **the time for one rotation is T**, then in the first quarter cycle from time t = 0 to t = ¼T, the AC emf and the current attain maximum value of emf V_{max} and of current I_{max}. In the next quarter cycle time ¼T to ½T, the two values fall back to zero. Then the polarity of AC emf reverses and the AC current flows in the opposite direction. It rises again to a maximum ($-V_{max}$) from time ½ T to ¾ T and falls to zero in the last quarter cycle, time ¾T to 2T (Figure 5.1). The cycle then repeats.

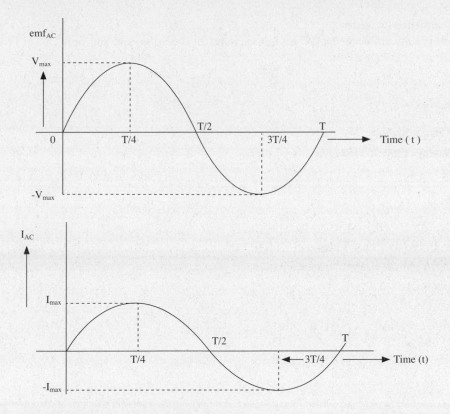

Fig. 5.1: The AC voltage and the current for one cycle.

The AC voltage V(t) at any time t is given by:

$$V(t) \quad = \quad V_{max} \sin\omega t$$

$$= \quad V_{max} \sin 2\pi f t$$

$$= \quad V_{max} \sin \frac{2\pi t}{T} \qquad \qquad \text{as } f = 1/T$$

The angle that the coil makes during its rotation is $\theta = \omega t$.

The AC current I(t) at any instant is

$$I(t) \quad = \quad \frac{V_{max}}{R} \sin\omega t$$

$$= \quad I_{max} \sin\omega t.$$

$$= \quad I_{max} \sin 2\pi f t \qquad \qquad \qquad 5.2$$

We can see that the maximum AC voltage occurs when $\sin\omega t$ is equal to its maximum value which is 1 when $\theta = 90°$ [$\pi/2$] or $270°$ [$3\pi/2$]. The minimum values occur at $\theta = 0°$ and $180°$ [π rad].

5.1.1 Instantaneous Value

The instantaneous value of the AC emf is the value of the voltage at any time t. It is given by

$$V(t) \quad = \quad V_{max} \sin 2\pi ft$$

And the current is
$$I(t) \quad = \quad I_{max} \sin 2\pi ft$$

5.1.2 AC Peak Value

The **maximum or peak value** V_{max} is given by

$$V_{max} \quad = \quad 2\pi f \, BAN$$

$$I_{max} \quad = \quad \frac{2\pi fBAN}{R}$$

5.1.3 The Peak-to-Peak Value

The **Peak-to-Peak (p-p) Value** is $2V$ max.

$$V_{p-p} \quad = \quad 2 \, V_{max} \qquad\qquad 5.3$$

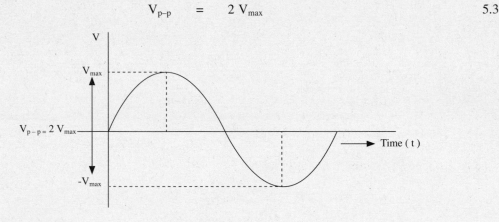

Fig. 5.2: The peak-to-peak (p-p) value of the AC voltage is $2V_{max}$.

5.1.4 Root Mean Square Value

The AC current and the voltage change continuously during a cycle. Both are sinusoidal and repeat after every cycle. To determine the amount of the current or voltage, the calculations would have to be carried out per cycle. The calculated average current per cycle is equal to zero. This does not present the true situation as the current was not zero during the cycle. The root-mean-square (rms) value of the current or the voltage is used to determine the AC magnitude.

   The root-mean-square (rms) value is the root of the average of the squares of current during a cycle.

   For current, the rms value is given by

$$I_{rms} \quad = \quad \sqrt{\frac{I_1^2 + I_2^2 + I_3^2 + \ldots\ldots I_n^2}{n}}$$

When this value is calculated, it is found to be equal to:

$$I_{rms} = \frac{I_{max}}{\sqrt{2}}$$ 5.4

$$= 0.707 \, I_{max}$$

The instruments that measure AC are calibrated in rms values.

The maximum value of current is:

$$I_{max} = 1.41 \, I_{rms}$$

Similarly $V_{rms} = 0.707 \, V_{max}$

and $V_{max} = 1.41 \, V_{rms}$ 5.5

Test yourself: An alternating current flows in a resistor of resistance 10 Ω. If heat is dissipated at a rate of 20 W, what is the rms value of the alternating current?

5.1.5 Phase

The angle θ determines the phase of alternating current or voltage. For instance, it can be seen in the following figure that when the angle is 90° ($\pi/2$), the current is maximum. The phase is $\pi/2$ rad. At a phase of 180° (π rad), the current is zero.

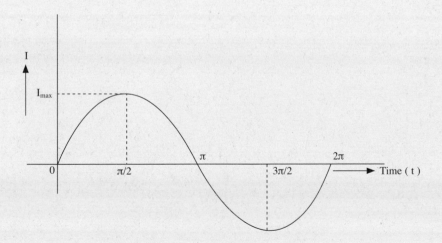

Fig. 5.3: Values of AC current at $\theta = \pi/2$, $\theta - \pi$, $\theta - 3\pi/2$ and $\theta = 2\pi$.

Two sinusoidal waves may be equal in all respects but different in phases as shown in Figure 5.4.

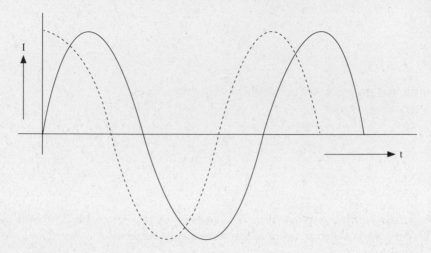

Fig. 5.4: Two equal AC currents at different phases.

When the bold line waveform is at zero, the dotted line waveform is maximum. After 90°, the bold line maximum is attained while the dotted line is zero. There is a 90° phase difference between the two waveforms.

Let us compare the alternating current (dotted) and alternating voltage (bold) graphs in the following figure.

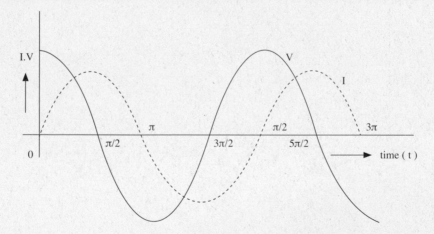

Fig. 5.5: The voltage V leads the current I by 90°.

The phase difference between the voltage and the current is $\pi/2$. At t = 0, the voltage is maximum and the current is zero. After $\pi/2$, the current is maximum and the voltage falls to zero. The voltage leads the current by an angle of 90° or $\pi/2$. There is a phase difference of 90° between the current and the voltage. In vector notation, if the voltage is taken along the positive x-axis, the current vector will be shown along the negative y-axis (Figure 5.6).

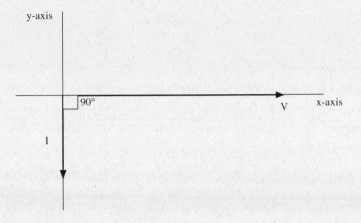

Fig. 5.6: The voltage V leads the current I by 90° (vector notation).

On the other hand, the following figure shows the voltage lagging the current by 90°.

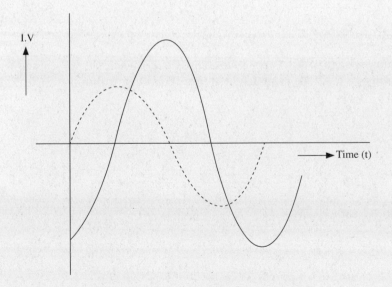

Fig. 5.7: The current lagging the voltage by 90°.

The vector diagram is given in Figure 5.8:

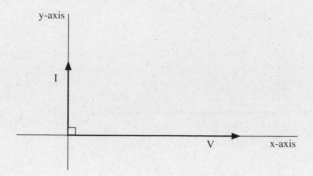

Fig. 5.8: The current I leads the voltage V by 90°.

5.2 AC Circuits

In this section, alternating current through resistors, capacitors, inductors and their basic combinations will be discussed. The behaviour of alternating currents is different from that of direct currents in the capacitors and the inductors.

5.3 AC through a Resistor

When an alternating voltage source is connected across a resistance, the AC current remains in phase with the voltage (Figure 5.9). The potential difference across the resistor is

$$V_{AC} \quad = \quad V_{max} \sin\omega t$$

The current is
$$I_{AC} \quad = \quad \frac{V_{max}}{R}$$

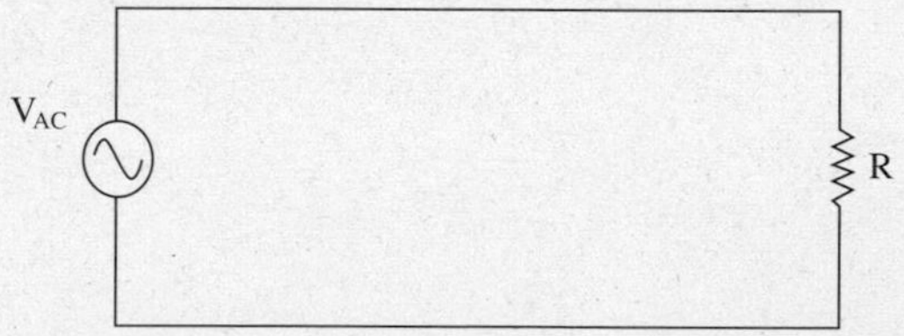

Fig. 5.9: Circuit with an AC power supply and resistance R.

This means that as the AC voltage increases, the current increases and when the voltage falls to zero and changes polarity, the current follows suit. The waveforms and vector diagram for AC through a resistor is shown in the following figure.

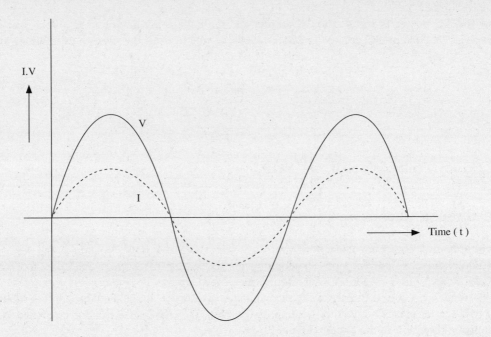

Fig. 5.10: In a purely resistive circuit, the current and the voltage are in phase.

In vector form, this can be shown as in Figure 5.11.

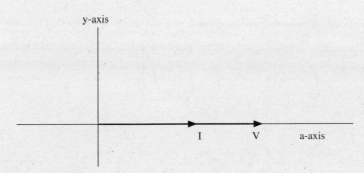

Fig. 5.11: AC voltage V and current I are in phase (vector picture).
V_{max}/R is the maximum AC current

$$\text{Current I} \quad = \quad I_{max} \sin 2\pi ft.$$

As the current flows through the resistance R, power loss takes place as was the case in DC.

AC Power Loss P	$=$	$V_{rms} I_{rms}$	5.6
	$=$	$I_{rms}^2 R$ as $V_{rms} = I_{rms} R$	5.7
	$=$	$\dfrac{V_{rms}^2}{R}$	5.8

The unit for AC power is watts. The AC current and the voltage are measured in amperes and volts respectively. In Pakistan, AC voltage is 220 V – 240 V, which gives the maximum value of voltage V_{max} as:

$$V_{max} \quad = \quad \sqrt{2} \; (V_{rms})$$

$$= \quad (1.41) \, (220)$$

$$= \quad 310.2 \text{ V}$$

The AC frequency in Pakistan is 50 c/s or 50 Hz.

5.4 AC through Capacitor

In DC circuits, the current in a capacitor flows for a very short time only. As soon as the capacitor is charged completely, the current stops. Alternating current on the other hand continues to flow in a capacitive circuit (with a capacitor only). During the first quarter cycle, the capacitor charges. As the voltage decreases in the next quarter cycle, the capacitor discharges. It again charges and discharges in the next two quarter cycles but with opposite polarity. Thus, the charges flow in one direction and then in the opposite direction in the circuit (Figure 5.12).

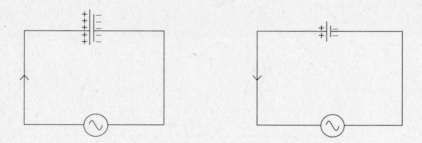

Fig. 5.12: Charging and discharging of a capacitor during one half AC cycle.

The voltage across a capacitor V_{rms} is given by

$$V_{rms} \quad = \quad I_{rms} \, X_C \qquad\qquad 5.9$$

where X_C is analogous to R in V = IR and is the opposition to the flow of current in capacitors. It is called the **capacitive reactance** and is inversely proportional to the frequency and capacitance of the capacitor:

$$X_c \quad = \quad \frac{1}{2\pi f C} \qquad\qquad 5.10$$

As AC frequency f increases, X_c decreases as shown in the following figure.

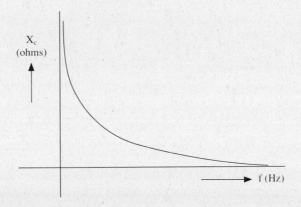

Fig. 5.13: X_c vs f in a capacitative circuit.

The capacitive reactance X_C becomes extremely high as the frequency approaches zero ($X_C = \infty$). At high frequencies, on the other hand, the capacitive circuit offers no opposition to the flow of current.

$$X_C \; \cong \; 0 \qquad \text{at high frequencies}$$

X_c is measured in ohms.

As

$$V(t) \;=\; V_{max} \sin 2\pi ft$$

$$\frac{dV}{dt} \;=\; V_{max}\, 2\pi f \cos 2\pi ft \qquad\qquad 5.11$$

and

$$Q \;=\; C\,V$$

$$\frac{dQ}{dt} \;=\; C\,\frac{dV}{dt}$$

$$I(t) \;=\; C\,\frac{dV}{dt}$$

Putting in value of dV/dt from equation 5.11, I(t) becomes

$$I(t) \;=\; C\,V_{max}\, 2\pi\, f \cos 2\pi f\, t$$

$$\;=\; V_{max}\, 2\pi fC \cos 2\pi f\, t$$

$$\;=\; \frac{V_{max} \cos 2\pi ft}{X_c} \qquad \text{using equation 5.10} \qquad\qquad 5.12$$

$$\;=\; I_{max} \cos 2\pi f\, t$$

Graphing V(t) and I(t) through a capacitor (Figure 5.14).

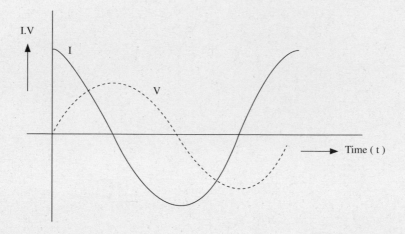

Fig. 5.14: AC voltage and current in a capacitive circuit.

In Figure 5.15, the vector diagram shows that the current leads the voltage in a capacitive circuit. The current reaches its maximum value a quarter-cycle ahead of the voltage reaching its maximum value.

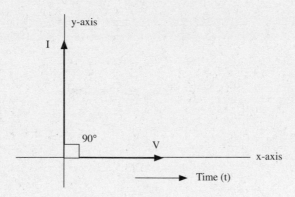

Fig. 5.15: I leads V in capacitive circuit in vector notation.

Power loss in capacitors is zero. During the first quarter cycle, V and I are positive. The energy from the power supply is used in charging the capacitor. In the next quarter cycle, V remains positive but I flows in the opposite direction — it is negative. Power is -VI. The energy is returned to the power supply. In the third quarter cycle, V and I both are negative. The energy is then stored in the capacitor as the other plate becomes positively charged. In the last quarter cycle, energy is restored again to the power supply (Figure 5.16). When added the total power loss per cycle is zero.

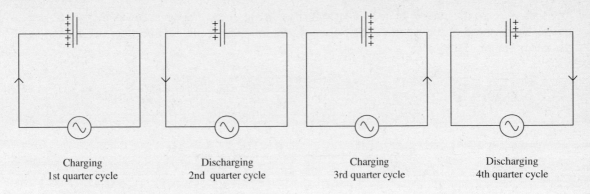

| Charging
1st quarter cycle | Discharging
2nd quarter cycle | Charging
3rd quarter cycle | Discharging
4th quarter cycle |

Fig. 5.16: Charging and discharging of capacitor.

5.5 AC through Inductors

Inductors are coils (solenoid coils). When the current flows through these coils, a magnetic field is generated.

In the case of AC, as the current rises, the magnetic field in the coil increases. In the next quarter cycle, the current falls and the magnetic field decreases. In the third and fourth quarter cycles, the current flows in the opposite direction so that the magnetic field also reverses its polarity. Due to the changing magnetic field, a back emf results. When the voltage is maximum, the current is zero. The current keeps rising whilst voltage decreases to zero. In the next quarter cycle, the voltage reverses and starts increasing to a maximum while the current drops to zero and so on in each succeeding AC cycle (Figure 5.17).

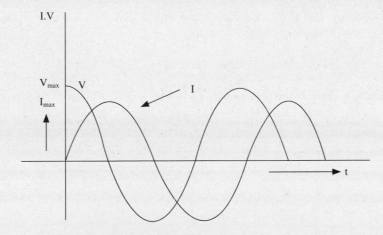

Fig. 5.17: Voltage and current through an inductor.

The voltage leads the current by $\pi/2$. In an inductive circuit, the voltage reaches its maximum value a quarter cycle ahead of the current. This can be shown in a phasor diagram keeping the voltage along the x-axis (Figure 5.18).

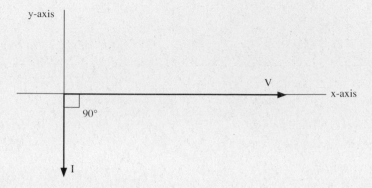

Fig. 5.18: Voltage leads the current in inductor (vector diagram).

The opposition to the flow of current in an inductor is called **inductive reactance X_L.** It is given by

$$V_{rms} = I_{rms} X_L \qquad 5.13$$

It is directly proportional to the frequency of the AC supply and the inductance of the inductor:

$$
\begin{aligned}
X_L &= 2\pi fL \\
&= \omega L \qquad 5.14
\end{aligned}
$$

where f is the frequency and L the self inductance of the coil. In contrast to capacitors inductive reactance increases with frequency of the AC supply.

$$X_L \quad \alpha \quad f \qquad 5.15$$

Taking the frequency f along the x-axis and X_L along the y-axis, a straight line between X_L and f is obtained (Figure 5.19).

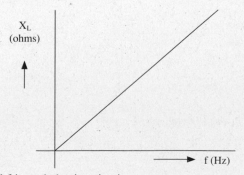

Fig. 5.19: Graph between X_L and f in an inductive circuit.

The relation between the current and the voltage can also be derived mathematically using the equation for self-inductance. The back emf is given by

$$V = -L \frac{dI}{dt}$$

Differentiating $I = I_{max} \sin\omega t$ 5.16

gives $V = \frac{dI}{dt} \omega I_{max} \cos\omega t$

gives $V = -L \omega I_{max} \cos\omega t$

$$= -\omega L \, I_{max} \sin(\omega t + \pi/2) \text{ as } \cos\omega t = \sin(\omega t + \pi/2)$$

Applying Kirchhoff's Law and using equation 5.14, the voltage V is equal to

$$V = X_L I_{max} \cos\omega t$$

$$= V_{max} \cos\omega t \qquad\qquad 5.17$$

Comparing equation 5.16 and 5.17, it can be seen that the voltage leads the current I by $\pi/2$ rad.

No power loss takes place in a purely inductive circuit. Energy is stored in the inductor and then sent back to the battery twice in each cycle with no power loss.

Test yourself: A current of 25 mA flows in an inductor of 10 H. If the AC frequency is 60 Hz, what is the potential difference across the inductor?

5.6 Impedance

The ratio of voltage to current in an AC circuit is called the Impedance. The resistance and reactance together offer the opposition to flow of current in AC circuits. **The total opposition to flow of current in AC circuits is called the Impedance.** It is the vector sum of the resistance, inductive reactance and capacitive reactance. The symbol for the impedance is **Z**.

$$\mathbf{Z} = \frac{V_{rms}}{I_{rms}} \qquad\qquad 5.18$$

Its unit is ohms.

5.7 R-C and R-L Series Circuits

5.7.1 R-C Series Circuit
A capacitor of capacitance C and resistor of resistance R are connected in series to the sinusoidal AC power supply (Figure 5.20). The same current flows through R and C as they are in series. The maximum voltage V_R across the resistor is in phase with I but the maximum voltage V_C across the capacitor will lag behind V_R by 90°. Keeping the current I as reference because the same current flows through R and C, the phasor diagram is constructed.

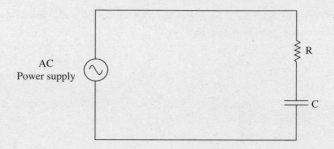

Fig. 5.20: An R-C Series Circuit.

Let the current I be along x-axis, then V_R will also be along the x-axis but V_C will be along negative y-axis (Figure 5.21).

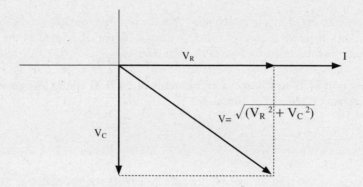

Fig. 5.21: Phasor diagram showing V_R and V_c and the resultant voltage V.

The resultant V can be determined using the Pythagoras Theorem.

$$V_{rms} = \sqrt{(V_R^2 + V_C^2)}$$

$$= \sqrt{(I_{rms}^2 R^2 + I_{rms}^2 X_C^2)}$$

$$= I_{rms} \sqrt{(R^2 + X_C^2)}$$

$$= I_{rms} \sqrt{\left[R^2 + \frac{1}{(\omega C)^2}\right]}$$

The impedance Z is equal to

$$Z = \frac{V_{rms}}{I_{rms}} = \sqrt{\left[R^2 + \frac{1}{(\omega C)^2}\right]} \qquad 5.19$$

The phase difference $-\theta$ between current I and voltage V in R-C series circuit is given by

$$\theta = \tan^{-1} \frac{V_C}{C_R}$$

$$= \tan^{-1} \frac{I_{rms} X_C}{I_{rms} R}$$

$$= \tan^{-1} \frac{X_C}{R}$$

$$= \tan^{-1} \frac{I}{\omega CR} \qquad\qquad 5.20$$

Example: Calculate the impedance, potential difference across a resistance R of 15 kΩ, potential difference across capacitor of capacitance $0.3\mu F$ and the phase shift in a R-C series circuit of frequency $\omega = 150$ rads^{-1} in which 1mA current is flowing.

$$R = 15k\Omega$$

$$C = 0.3\mu F$$

$$\omega = 150 \text{ rads}^{-1}$$

$$I = 1 \text{ mA}$$

$$V_R = IR$$

$$= (1 \times 10^{-3} \text{ A}) (15 \times 10^3 \ \Omega)$$

$$= 15 \text{ V}$$

$$V_C = I_{rms} X_C$$

$$= I_{rms} [1/\omega C]$$

$$= (1 \times 10^{-3} \text{ A}) [1/(150 \text{ rads}^{-1}) (0.3 \times 10^{-6} \text{ F})]$$

$$= \frac{1}{45} \times 10^3$$

$$= 22.22 \text{ V}$$

The resultant voltage V is given by

$$V = \sqrt{(V_R{}^2 + V_C{}^2)}$$

$$= \sqrt{15^2 + 22.22^2}$$

$$= \sqrt{718.72}$$

$$= 26.81 \text{ V}$$

Example: (contd.)

Impedance equals	Z =	$\dfrac{V_{rms}}{I_{rms}}$
	=	$\dfrac{26.81V}{1\times 10^{-3}}$
	=	$26.81 \times 10^3\ \Omega$
Phase Shift	θ =	$\tan^{-1} \dfrac{V_C}{V_R}$
	=	$\tan^{-1} \dfrac{22.22}{15}$
	=	$55.97°$

5.7.2 R-L Series Circuit

Fig. 5.22: An inductor L and resistance R is connected in series with an AC power supply.

As it is shown in Figure 5.22, resistance R and an inductor of inductance L are connected in series with the AC voltage supply. The current I again flows through both the circuit elements, that is, the inductor and resistor, so it can be considered as reference. Taking current I along the x-direction, V_R would also be along this direction while V_L would lead the current by 90°. Its direction is along the positive y-axis (Figure 5.23).

Here
$$V_R = IR$$
$$V_L = IX_L$$

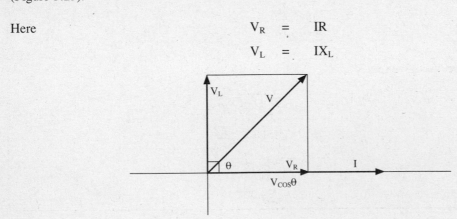

Fig. 5.23: Resultant V, V_L and V_R in R-L Series Circuit.

The resultant voltage is given by

$$V = \sqrt{(V_L{}^2 + V_R{}^2)}$$

$$= I\sqrt{(X_L{}^2 + R^2)}$$

$$= I\sqrt{[(\omega L)^2 + R^2)]}$$

The impedance Z is given by

$$Z = \frac{V}{I}$$

$$= \sqrt{[(\omega L)^2 + R^2]} \qquad 5.21$$

5.8 Power in AC Circuits

As the current and the voltage are in phase in resistors, instantaneous power dissipated is

$$P = I(t)\, V(t) \qquad 5.22$$

where I(t) and V(t) are instantaneous current and voltage respectively

$$P = (I_o \sin\omega t)(V_o \sin\omega t)$$

$$= I_o V_o \sin^2 \omega t$$

where I_o and V_o are maximum current and voltage.

$$= I_o{}^2 R \sin^2\omega t. \qquad \text{As } V_o = I_o R \qquad 5.23$$

The maximum power is when sin ωt is 1 or $\omega t = 90°$

$$P_{max} = I_o{}^2 R$$

$$= I_o V_o \qquad 5.24$$

The maximum power is the product of maximum values of I and V.

But $\qquad I_{rms} = \dfrac{I_o}{\sqrt{2}} \quad$ and $\quad V_{rms} = \dfrac{V_o}{\sqrt{2}}$

The average power is given by

$$P_{av} = I_{rms}\, V_{rms.} \qquad 5.25$$

As the current and the voltage are 90° out of phase in purely capacitive or inductive circuits, no energy is dissipated in these circuits. In other words, there is no power loss. Power loss only occurs in a resistance R where the voltage and current are in phase. A resistance R is connected in the circuit (R-C, R-L or R-L-C circuits) and the resultant voltage V (Figures 5.21, 5.23 and 5.26) has a phase equal to θ. Power loss would be the product of the current I_{rms} and the component of the resultant voltage in the direction of current I_{rms}. If the voltage across the resistance R is taken along the x-axis, the component of the resultant voltage V in the direction of V_R along the x-axis would contribute to the power loss. In other words, the power loss in these circuits is the component of resultant voltage in phase with the current.

Thus, the power loss would be

$$\text{Power loss} \quad P \quad = \quad (V_{rms}\cos\theta) \, I_{rms}$$

$$= \quad I_{rms} V_{rms} \cos\theta \qquad\qquad 5.26$$

In equation 5.26, **cosθ is known as the power factor.**

No power is dissipated in an inductor or capacitor. Power loss is restricted to resistors.

When R = 0 and θ = ± 90°, $P_{av} = 0$ and the circuit is a purely inductive or capacitive circuit.

Power is measured in watts in SI system of units.

Test yourself: What is the angle of the resultant voltage if the power factor is 0.87?

Example: An AC heater of 1000 W is connected to a voltage source of 240 V_{rms}. What is the resistance of the heater and the maximum current through it?

$$P_{av} \quad = \quad 1000 \text{ W}$$

$$V_{rms} \quad = \quad 240 \text{ V}$$

$$P_{av} \quad = \quad I_{rms} V_{rms}$$

$$= \quad V_{rms}^2/R$$

$$R \quad = \quad \frac{V_{rms}^2}{P_{av}}$$

$$= \quad \frac{(240)^2}{1000}$$

$$= \quad 57.6 \ \Omega$$

$$I_{rms} \quad = \quad \frac{P_{av}}{V_{rms}}$$

$$= \quad \frac{1000}{240}$$

$$= \quad 4.167 \text{ A}$$

$$I_{rms} \quad = \quad \frac{I_{max}}{\sqrt{2}}$$

$$I_{max} \quad = \quad \sqrt{2} \, I_{rms}$$

$$= \quad 5.89 \text{A}$$

As mentioned earlier, no power is dissipated in a capacitor as the current and voltage are 90° out of phase. When charging both the current and the voltage are positive. Energy is stored in the capacitor. In the next quarter cycle, the voltage is positive but the current is negative. The capacitor is discharging and the energy is sent back to the generator.

In the next half cycle, this is repeated except that polarities are changed. Energy is sent back and forth between capacitor and power supply. No power is dissipated; energy is stored in the electric field of the capacitor.

The same is true for inductors. The current and voltage are 90° out of phase. Energy is sent back and forth between inductor and power supply. Energy is stored in the magnetic field of the inductor.

5.9 Series Resonance Circuit

A resistor R, capacitor C and inductor L are connected in series across an AC power supply. This is termed as R-L-C Series Circuit. Let the voltages be V_R, V_C and V_L across the resistance, capacitor and inductor respectively. The same current I flows through all three as they are connected in series (Figure 5.24).

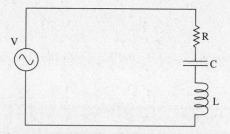

Fig. 5.24: An R-L-C Series Circuit.

Using the same procedure as in the previous two sections, V_L leads I by 90° while V_C lags behind I by 90° and V_R is in phase with I. The phasor diagram is shown in Figure 5.25.

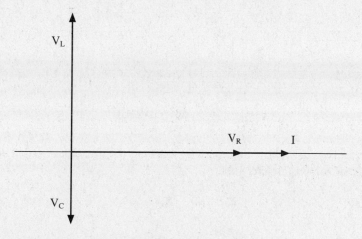

Fig. 5.25: Phasor diagram for V_R, V_C and V_L in a R-L-C Series Circuit.

Adding V_C and V_L vectorally gives $V_L - V_C$ and so the phasor diagram reduces as shown in Figure 5.26.

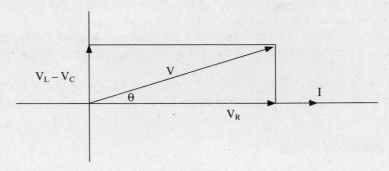

Fig. 5.26: Phasor diagram between $(V_L - V_C)$ and V_R.

Applying Pythagoras Theorem to determine the relation between the applied voltage V and V_R, V_L and V_C

$$V = \sqrt{[V_R{}^2 + (V_L - V_C)^2]}$$

Here the voltages are for maximum values. By dividing both the sides of the equation by $\sqrt{2}$, the rms values are obtained.

$$V_{rms} = I_{rms}\sqrt{\left[R^2 + \left\{\omega L - \frac{1}{\omega C}\right\}^2\right]}$$

The impedance Z is

$$Z = \frac{V_{rms}}{I_{rms}}$$

$$= \sqrt{\left[R^2 + \left\{\omega L - \frac{1}{\omega C}\right\}^2\right]} \qquad 5.27$$

The phase angle θ is

$$\theta = \tan^{-1}\left(\frac{X_L - X_C}{R}\right)$$

$$\theta = \tan^{-1}\frac{\omega L - 1/(\omega C)}{R} \qquad 5.28$$

The impedance Z can be reduced to a minimum by reducing the numerator to 0. This can be achieved at a particular value of $\omega = \omega_o$. At this value

$$X_L = X_C$$

$$\omega_o L = \frac{1}{\omega_o C}$$

$$\omega_o{}^2 = \frac{1}{LC} \qquad\qquad 5.29$$

$$\omega_o = \frac{1}{\sqrt{LC}}$$

ω_o is the angular resonant frequency. As $\omega_o = 2\pi f_o$, equation 5.29 can be written as

$$4\pi^2 f_o{}^2 = \frac{1}{LC}$$

$$f_o{}^2 = \frac{1}{4\pi^2 LC}$$

$$\mathbf{f_o} = \frac{\mathbf{1}}{\mathbf{2\pi\sqrt{LC}}} \qquad\qquad 5.30$$

Thus, at a particular frequency, **f_o which is called the resonance frequency**, the impedance is minimum. At this frequency, the current will be maximum.

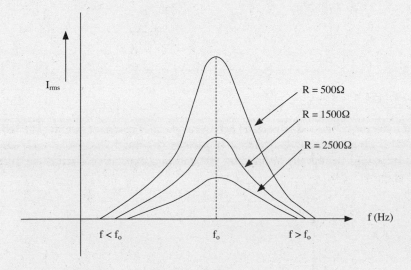

Fig. 5.27: Graph between frequency f and current for an R-L-C Series Circuit.

A graph between frequency f and current I for different resistances is shown in Figure 5.27. Thus, it is seen at resonance [$f_o = 1/(2\pi\sqrt{LC})$], **the capacitive reactance is equal to the inductive reactance.**

$$X_C = X_L \qquad\qquad 5.31$$

The circuit behaves as a **purely resistive circuit**. This happens at the resonance frequency f_o. **At this frequency, the current becomes maximum and the impedance becomes minimum which is equal to R.**

$$Z = \sqrt{[R^2 + (X_L - X_C)^2]}$$

$$= \sqrt{(R^2 + 0)}$$

$$= R \qquad\qquad 5.32$$

It is observed that the peak values increase as resistance decreases which is expected.

5.10 Parallel Resonance Circuit

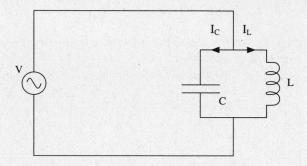

Fig. 5.28: A Parallel R-L-C Circuit.

A capacitor and inductor connected in parallel behave as parallel resonant circuit. The inductor coil has a resistance R. This circuit is an R-L-C circuit in parallel (Figure 5.28). We already know that in an inductor, the current lags behind the voltage by 90° and in a capacitive circuit, the current leads the voltage by 90° (Figure 5.29).

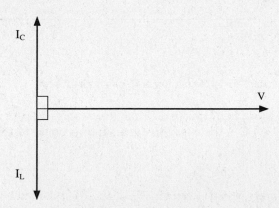

Fig. 5.29: Phasor diagram for R-L-C Parallel Circuit.

In the circuit, the voltage V is taken as reference as the same potential difference is across the inductor and capacitor

Then

$$V = I_C X_C \quad \text{and} \quad I_C = \frac{V}{X_C}$$

$$V = I_L X_L \quad \text{and} \quad I_L = \frac{V}{X_L}$$

The net current is given by

$$I = I_C - I_L$$

The current I can be made equal to zero at a particular frequency given by $\omega = \omega_o$.

At resonance

$$I = 0$$

$$I_C - I_L = 0$$

$$I_C = I_L$$

$$V X_C = V X_L \qquad \text{as } I = V/R$$

$$V(\omega_o C) = \frac{V}{\omega_o L}$$

$$\omega_o C = \frac{1}{\omega_o L}$$

$$\omega_o^2 = \frac{1}{LC}$$

$$f_o = \frac{1}{2\pi\sqrt{LC}}$$

As any circuit also has a resistance R, a current I_R will be flowing at resonance when I_C and I_L cancel each other.

In parallel circuits at resonance, the current is minimum and the impedance maximum (Figure 5.30). The current and the voltage are in phase.

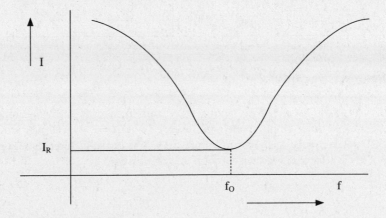

Fig. 5.30: I vs f in an R-L-C Parallel Circuit.

The parallel resonance circuit can be used in radio receivers. As the capacitance C is altered, the resonance frequency changes and signals of different frequencies can be detected.

Example: A variable capacitor [C = 100 pF to 600pF] is connected in parallel to an inductance of 200 μH. Calculate the range of frequencies that the circuit can detect.

$$C \quad = \quad 100pF - 600pF$$

$$L \quad = \quad 200\mu H$$

$$f_1 \quad = \quad \frac{1}{2\pi\sqrt{(LC)}}$$

$$= \quad \frac{1}{2\pi\sqrt{(100\times 10^{-12})(200\times 10^{-6})}}$$

$$= \quad \frac{10^7}{(2\pi)(1.414)}$$

$$= \quad 0.1126 \times 10^7$$

$$= \quad 1.126 \times 10^6 \text{ Hz}$$

$$f_2 \quad = \quad \frac{1}{2\pi\sqrt{(LC)}}$$

$$= \quad \frac{1}{2\pi\sqrt{(600\times 10^{-12})(200\times 10^{-6})}}$$

$$= \quad \frac{10^7}{2\pi\sqrt{12}}$$

$$= \quad \frac{10^7}{2\pi\,(3.464)}$$

$$= \quad 0.0459 \times 10^7$$

$$= \quad 4.5 \times 10^5 \text{ Hz}$$

The frequency range is 4.5×10^5 Hz to 1.126×10^6 Hz

5.11 Three Phase AC Supply

A three phase AC voltage is produced by having three coils placed at 120° to each other in an AC generator. There would be three outputs, one from each coil. At t =0, the outputs would have phases of 0°, 120° and 240° respectively as shown in Figure 5.31. The phase difference between the first two outputs would be 120° whilst that between the first and third would be 240°.

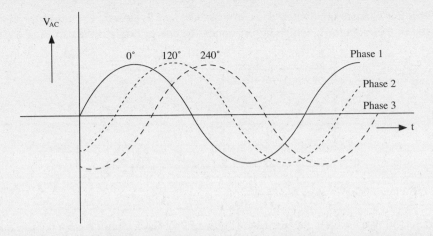

Fig. 5.31: Three phases of AC voltage.

One output from each of the slip rings is grounded and so the external circuit has four terminals, one at 0 volts and the other three at 240 volts with respect to the grounded terminal. The external connections A, B, C and D are shown in Figure 5.32. D is grounded and is at 0V.

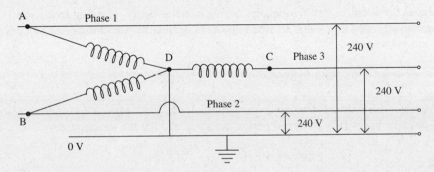

Fig. 5.32: External connections for three phase AC supply.

The three phase AC is very useful because power is distributed. Some appliances can be placed across one phase. Lights and fans can be across another phase and air conditioners along a third phase. You must have noticed that at times some of your home appliances on one phase are off while the other phases are on. Had all the appliances been placed on one phase, the power line may get loaded and the voltage would drop. In such cases, the circuit breakers would shut off the heavy load appliances like air conditioners.

5.12 Principle of Metal Detectors

A circuit with an inductor and capacitor in parallel can behave like an oscillator (Figure 5.33). This is similar to a pendulum that changes kinetic energy to potential energy repeatedly as it oscillates to and fro. It would continue to do so indefinitely if no energy is lost. In the L-C circuit, when switch S_1 is closed, the capacitor charges up. Switch S_1 is now opened and S_2 closed. The capacitor discharges and energy is stored in the inductor L, which in turn sends the energy back to the capacitor C. The process continues.

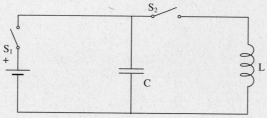

Fig. 5.33: An L-C circuit behaves as an oscillator.

In the metal detector, two such circuits are used (Figure 5.34). Each circuit has its own frequency given by

$$f_1 = \frac{1}{2\pi\sqrt{L_1 C}}$$

$$f_2 = \frac{1}{2\pi\sqrt{L_2 C}}$$

Small differences between the two frequencies are detected as **beats.** The beat frequency is $f_1 - f_2$. The frequencies are so adjusted that when no metal object is nearby, the two frequencies f_1 and f_2 are equal and so there are no beats:

$$\text{Beat frequency} = f_1 - f_2 = 0$$

When a metal object is brought near oscillating circuit 1, L_1 decreases and f_1 increases to f_1'. A beat frequency is produced which is amplified and is detected through the headphones:

$$\text{Beat frequency} = f_1' - f_2$$

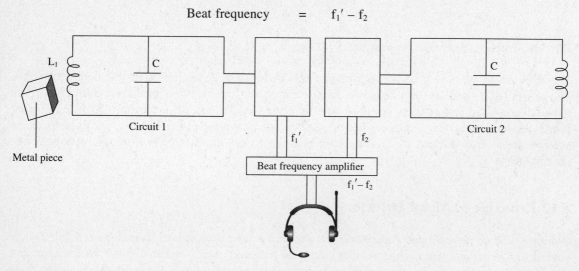

Fig. 5.34: A schematic diagram for a metal detector.

A beat frequency means a metal object is detected. These days metal detectors are used in checking bags for concealed guns, etc. in hotels and bank security systems.

Example: The inductance in the search coil of a metal detector changes from 100 μH to 101μH when a metal is brought near it. The resonance frequency for both detectors is 900 kHz. What is the beat frequency?

$$L_1 \;=\; L_2 \;=\; 100 \ \mu H$$

$$L' \;=\; 101 \ \mu H$$

$$f_1 \;=\; f_2 \;=\; 900 \text{ kHz}$$

$$\frac{f_1'}{f_2} \;=\; \frac{1}{2\pi\sqrt{L_1'C}} \;\div\; \frac{1}{2\pi\sqrt{L_2C}}$$

$$= \frac{\sqrt{L_2}}{\sqrt{L_1'}}$$

$$= \sqrt{\frac{100}{101}}$$

$$= 0.995$$

Now
$$f_1' \;=\; f_2\,(0.995)$$

$$= (900)\,(0.995)$$

$$= 895.5 \text{ Hz}$$

Beat frequency
$$= f_1' - f_2$$

$$= 900 \text{ kHz} - 895.5 \text{ kHz}$$

$$= 4.5 \text{ kHz}$$

5.13 Choke

A choke consists of a coiled copper wire around an iron core (Figure 5.35). The inductance of such a coil is high but its resistance is low. The current is reduced by the high inductance L and very little power is dissipated as resistance R is small. Choke coils are placed in AC circuits to reduce/control the AC currents.

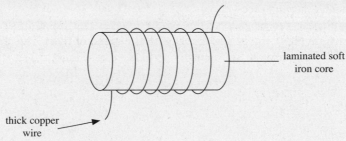

laminated soft iron core

thick copper wire

Fig. 5.35: Choke.

5.14 Electromagnetic Waves

We have seen that an electric current produces a steady magnetic field. And a changing magnetic flux produces an electric current.

The two effects do not seem to be exactly similar. Let us examine the charging of two metal plates. Supposing we have two metallic plates connected to a battery. As the plates charge up, an electric field will be produced between the plates.

During the time that the plates are being charged, there is an increasing (changing) electric field between the plates. A magnetic field which increases is produced around the electric field just as if a current was flowing. When the plates are charged up completely and the electric field is now constant, the magnetic field disappears.

When the plates are discharged and the electric field starts decreasing, a magnetic field appears again but in the opposite direction (Figure 5.36).

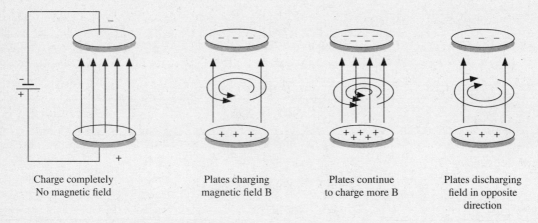

Charge completely
No magnetic field

Plates charging
magnetic field B

Plates continue
to charge more B

Plates discharging
field in opposite
direction

Fig. 5.36: Magnetic field associated with a changing electric field.

A changing magnetic field generates an electric field and a **changing electric field** generates a magnetic field.

If a changing magnetic field increases or decreases its rate of change, the electric field which is produced would also increase or decrease. In other words, a changing electric field can generate a changing magnetic field which in turn can produce a changing electric field. The two fields electric and magnetic generate each other in space.

These oscillating electric and magnetic fields produce waves that travel in straight lines in space. They are called **electromagnetic waves and travel with the speed of light**. The electric field, magnetic field and the direction of propagation are mutually perpendicular (Figure 5.37).

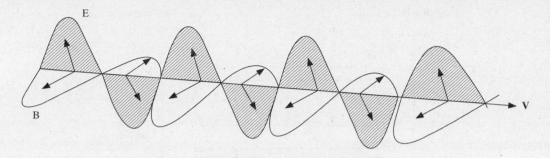

Fig. 5.37: The electric field E, magnetic field B and direction of propagation of the waves are mutually perpendicular.

5.14.1 Electromagnetic Spectrum

Unlike sound and water waves, electromagnetic waves do not require a material medium.

Light waves are electromagnetic waves. Although all electromagnetic waves move with the velocity of light c, their wavelength and frequencies vary.

The continuous band of wavelengths for the entire range of electromagnetic waves is called the electromagnetic spectrum. There are no sharp, definite boundaries between the different types of electromagnetic waves. Each section blends into the next and there is overlapping.

Radio waves have the longest wavelength and the smallest frequencies. They are produced by charges accelerating through conducting wires. They are used in radio and television communication systems. Microwaves (short-wavelength radio waves) have wavelengths ranging between about 1mm and 30cm and are generated by electronic devices. They are used in microwave ovens and radar systems used in aircraft navigation.

Infrared follow the microwave. Infrared as the name suggests have frequencies smaller than those of red light. Our eyes are only sensitive to the light frequencies. So we can feel the effects of the infrared radiation but not see them. They are produced by hot objects and molecules, and have the wavelength ranging from about 1mm to the longest wavelength of visible light, 7×10^{-7} m. They are readily absorbed by most materials. The infrared energy agitates the atoms of the object, increasing the vibrations of atoms.

Next is the visible spectrum starting from red colour and going to violet progressively with smaller and smaller wavelengths and increasing frequencies. Visible light waves have frequencies ranging from 4×10^{14} to 7.5×10^{14} c/s. Ultraviolet (UV) light wavelengths range from about 4×10^{-7} m (0.4 μm) down to 6×10^{-7} m (0.6 μm). The Sun is an important source of ultraviolet light (which is the main cause of suntans).

Beyond light are the ultraviolet, x-rays and γ-rays all dangerous to living things because of their penetrating power (smaller wavelengths) and energy carrying capacities (higher frequencies). Infrared, visible light and ultraviolet radiations are produced when an electron in an atom falls from a higher to a lower level.

The electromagnetic waves sometimes overlap. To determine the type of wave, their characteristics or the origin of the waves has to be found. X-rays are produced by highly accelerated electrons, while gamma rays are produced by radioactive nuclei. The electromagnetic spectrum is shown in Figure 5.38.

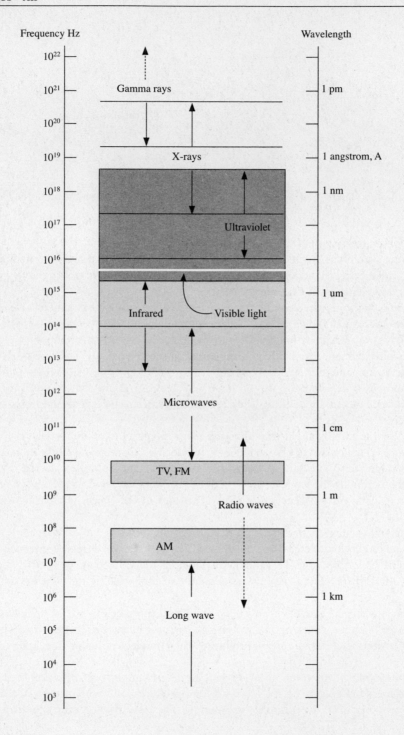

Fig. 5.38: The Electromagnetic Spectrum.

Test yourself: What is the wavelength of radio wave that oscillates with a frequency of 1500 kHz?

5.15 Principle of Generation, Transmission and Reception of Electromagnetic Waves

Electromagnetic waves play an important role in our daily lives. Radio and television programmes can be enjoyed because these waves can carry sound and picture signals.

Electromagnetic waves are produced when the magnetic flux changes. The magnetic flux changes when currents change. A current produces a magnetic field but unless the current increases or decreases, the field remains constant and no electromagnetic waves are produced. A changing current means accelerating (or decelerating) charge which produces electromagnetic waves. This is the **principle of radio transmission.**

At the radio station, charges (electrons) are made to oscillate in an aerial or antenna wire. In this way, sound and picture information can be transmitted over long distances. Antenna wires are connected through transformers to an AC power supply whose frequency can be altered by the inductor and variable capacitor connected to the circuit. The frequency is given by

$$f = \frac{1}{2\pi\sqrt{LC}}$$

Figure 5.39 shows the schematic diagram of a radio transmitter

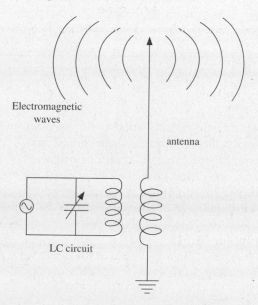

Fig. 5.39: A radio wave transmitter.

As the voltage changes from zero to maximum and back to zero in the first half cycle, charges in the metal antenna are first accelerated downwards and then decelerated. Then in the negative half cycle, the charges are accelerated upwards and then slowed down. The cycle repeats. As the charges accelerate and decelerate, energy is radiated out in the form of electromagnetic waves whose frequency matches that of the AC circuit. Different stations send there programmes using different frequencies. The waves are propagated with the speed of light c.

Radio receivers in our homes are L-C circuits which are tuned (by changing the capacitance C) to receive a particular frequency. As the frequency of the electromagnetic waves and the receiver circuit match, resonance occurs. A frequency is selected by the radio receiver. It is then amplified, and changed back to sound or picture for our viewing (Figure 5.40).

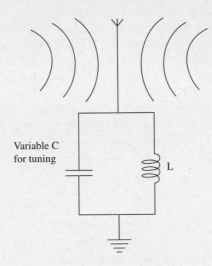

Fig. 5.40: Radio receiver.

5.16 Modulation

The process by which sound and light data is carried by the electromagnetic waves from the radio and television stations to receivers in the home, etc. is called modulation.

The signals to be transmitted, for example, music (sound waves), are first converted into low frequency signals and are then superimposed on the carrier electromagnetic waves.

In modulation, the **frequency or amplitude** of the carrier wave is changed by superimposing the light (from picture) or sound (from music or speech) signal on the carrier waves.

5.16.1 Amplitude Modulation (A.M)
Amplitude Modulation involves variation in amplitude of the carrier wave due to the changing amplitude of the superimposed signals (Figure 5.41).

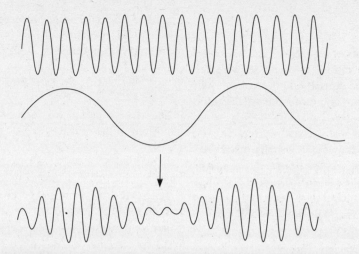

Fig. 5.41: Amplitude modulation.

In amplitude modulation, the radio frequencies range from small frequencies (540 Hz) to medium frequencies (1600 kHz).

5.16.2 Frequency Modulation

Frequency modulation changes the frequency of the carrier wave according to the frequency changes of the audio wave. These waves have a higher frequency (88 MHz to 108 MHz) range as compared to amplitude modulation waves.

The modulated wave has a higher frequency where the radio wave has high amplitude. Its frequency decreases with amplitude. It is minimum where amplitude is $-V_{max}$. The carrier frequency remains unaltered where the signal amplitude is zero. Frequency modulated waves have higher energies and are so less affected by outside disturbances but their range is shorter.

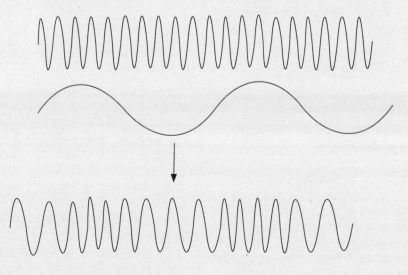

Fig. 5.42: Frequency modulated carrier wave.

Summary

If AC circuit consists of a generator and a resistor, the current in the circuit is in phase with the voltage. That is, the current and the voltage will attain their maximum and minimum values at the same time.

In discussion of voltage and current in AC, **rms values** of voltages are usually used. AC ammeters and voltmeters are designed to read rms values. The rms values of the current and voltage (I_{rms} and V_{rms}) are related to the maximum values of these quantities (I_{max} and V_{max}) as follows:

$$I_{rms} = \frac{I_{max}}{\sqrt{2}}$$

and

$$V_{rms} = \frac{V_{rms}}{\sqrt{2}}$$

The rms voltage across a resistor is related to the rms current in the resistor by **Ohm's Law:**

$$V_{rms} = I_{rms}R$$

If an AC circuit consists of a voltage source and a capacitor, the voltage lags behind the current by 90°.

The impeding effect of a capacitor C on current in an AC circuit is given by the **capacitive reactance Xc:**

$$Xc = \frac{1}{2\pi f C}$$

where f is the frequency of the AC voltage source.

The rms voltage across and the rms current in a capacitor are related by

$$V_{rms} = I_{rms}Xc$$

If an AC circuit consists of a voltage source and an inductor, the voltage leads current by 90°. That is, the voltage reaches its maximum value one quarter of a period before the current reaches its maximum values.

The effective impedance of a coil in an AC circuit is measured by a quantity called the **inductive reactance** X_L defined as

$$X_L = 2\pi f L$$

The rms voltage across coil is related to the rms current in the coil by

$$V_{rms} = I_{rms} X_L$$

In an **R-L-C series AC circuit**, the applied voltage V is related to the maximum voltages across the resistor V_R, capacitor V_C, and inductor V_L by

$$V = \sqrt{[V_R{}^2 + (V_L - V_C)^2]}$$

If an AC series circuit contains a resistor, an inductor, and a capacitor connected in series, the **impedance Z** of the circuit is defined as

$$Z = \sqrt{[R^2 + (X_L - X_C)^2]}$$

In an R-L-C series AC circuit, the applied rms voltage and the current are out of phase. The **phase angle** θ between the current and voltage is given by

$$\tan \theta = \frac{X_L - X_C}{R}$$

The resonance frequency in a R-L-C series circuit is given by

$$f_o = \frac{1}{2\pi\sqrt{LC}}$$

The **average power** delivered by the voltage source in an R-L-C series AC circuit is

$$P_{av} = V_{rms} I_{rms} \cos \theta$$

where the constant $\cos\theta$ is called the **power factor.**

The resonance frequency in a R-L-C parallel circuit is given by

$$f_o = \frac{1}{2\pi\sqrt{LC}}$$

In this case, impedance is maximum and the current is minimum.

Electromagnetic waves have the following properties:

They are transverse waves, because the electric and magnetic fields are perpendicular to the direction of travel. Electromagnetic waves travel with the speed of light.

Carrier waves are used in the transmission of sound and picture signals. Amplitude modulation can be transmitted over longer distances while frequency modulated waves are transmitted over shorter distances but are less affected by outside disturbances.

Questions

1. Distinguish between period and frequency of an alternating current.
2. What is the shape of a sinusoidal wave?
3. What do we mean by effective value of an alternating current? What is the other name for the effective current?
4. What do we mean when we say alternating current and voltage are in phase?
5. If the capacitance and inductance in an R-L-C series circuit are each doubled but resistance remained the same, what happens to the frequency?
6. Show that the inductive reactance X_L has SI units of ohms.
7. What causes the inductive reactance?
8. Inductors and capacitors behave differently when connected to AC power. How is their behaviour different?
9. Inductors and capacitors behave differently with AC and DC. How?
10. What is the relation between the reactance and frequency in a capacitor? What would be the reactance if frequency is zero (DC)?
11. What is the relation between the inductive reactance and frequency? What would be the

reactance when frequency is zero? How does an inductor behave when DC flows through it?
12. What is the difference between inductive reactance and capacitive reactance? Discuss impedance.
13. An R-L-C circuit is brought to resonance. a) What is the impedance of the circuit at resonance? b) When will the current be greatest: at resonance, at frequency less than the resonant frequency or at a frequency above the resonant frequency?
14. Is it possible to have an electric circuit with no resistance? Explain.
15. What is the importance of a choke coil?
16. What is the difference between x-ray and gamma rays?
17. Which has the highest frequency: radio waves, gamma waves, microwaves or visible light?
18. How fast do x-rays move in vacuum?
19. How can you differentiate between x-rays and gamma rays if both have the same wavelength and frequency?
20. How do you tune to different stations on your radios and television sets?
21. Distinguish between amplitude modulation (AM) and frequency modulation (FM)?

Problems

1. The peak value of AC is given by

    a) V_{rms}
    b) $\dfrac{V_{rms}}{\sqrt{2}}$
    c) $\sqrt{2\,V_{rms}}$
    d) $2\sqrt{V_{rms}}$

2. Find the effective value of an alternating current whose peak value is 10A.
3. An AC voltage has an output of $\Delta V = (200V)\sin 2\pi ft$. This source is connected to a 100 Ω resistor. Find the rms current in the resistor.

4. A capacitor is found to offer 30 Ω capacitive reactance when connected in a 500 Hz circuit. What is its capacitance?

5. A 150 F capacitor is connected to an alternating potential difference of 15 V and frequency 50 Hz. Calculate a) the capacitive reactance and b) the current in the circuit?

6. A coil has an inductance of 100 mH. What is its inductive reactance when used with a 60 Hz alternating current?

7. In a purely inductive AC circuit L = 25 mH and the rms voltage is 150 V. Find the inductive reactance and rms current in the circuit if the frequency is 60 Hz.

8. A coil with 60 Ω inductive reactance in series with a resistance of 25 Ω is connected to a 130 V AC line. What will be the current through the circuit?

9. An inductor has a 60 Ω reactance at 50 Hz. What will be the maximum current if this inductor is connected to a 60 Hz source that produces a 100 V rms voltage?

10. When a 4 μF capacitor is connected to a generator whose rms output is 50 V, the current in the circuit is 0.5 A. What is the frequency of the source?

11. The primary coil of a certain transformer has an inductive reactance of 1200 Ω when connected to a 60 Hz AC power line. What will be its inductive reactance when connected to a 25 Hz power line?

12. A coil with 60 Ω inductive reactance in series with a resistance in series with a resistance of 40 Ω, is connected to a 240 V AC power supply. What will be the current through the circuit?

13. An AC voltage V = (100V) sin (1000t) is connected to a series R-L-C circuit. If R = 400 Ω, C = .5 μF and L = 0.5 H, find the average power delivered to the circuit.

14. Find the impedance of a series R-L-C AC circuit for which R = 250 Ω, L = 0.6 H, C = 3.5 μF, f = 60 Hz and V_{max} = 150V.

15. A coil having 500 mH inductance and 100 Ω resistance is connected in series with a capacitor of 10 μF to a 220 V, 50 Hz line. What is the voltage across the coil?

16. An alternating current is one-half wavelength out of phase with its voltage. Graph the voltage and current.

17. When an AC current of 0.5 A flows in an inductor, the rate of heat dissipated is 15 W. The reactance of the inductor is 40 Ω. Find the impedance.

18. A 220 V AC motor draws a 5.5 A current when running at full power. If its power factor is 0.87, at what rate does it consume power?

19. A resonant circuit in a radio receiver is tuned to a certain station when the inductor has a value of 0.2 mH and the capacitor has a value of 30 pF. Find the frequency of the radio station and wavelength sent out by the station.

20. An inductor of inductance 100 H is connected in parallel with a variable capacitor whose capacitance can be varied from 300 pF to 50 pF. Calculate the frequency range of the circuit.

21. The AM band extends from approximately 500 kH to 1600 kHz. If a 2 μH inductor is used in a tuning circuit for a radio, what is the capacitance range for the capacitor?

22. What are the FM radio band (88 – 108 MHz) wavelength ranges?

23. What are the AM radio band (540 – 1600 kHz) wavelength ranges?

24. Red light with a frequency of 5×10^{14} Hz travels through space with a speed of 3×10^8 m/s. What is its wavelength?

25. A light wave of frequency of 6×10^{14} Hz appears green to the eye. What is its wavelength in cm?

6 Physics of Solids

OBJECTIVES

After studying the chapter, the student shall be able:
- to understand the difference between the crystalline and amorphous solids.
- to understand lattice and unit in crystalline structures.
- to understand what polymers are.
- to understand the stress resulting in strain.
- to understand the types of strains and moduli of elasticity, shear and bulk.
- to understand the band theory and based on the theory why some elements are conductors while others are insulators or semiconductors.
- to differentiate between intrinsic and extrinsic semiconductors.
- to understand superconductors and their importance.
- to distinguish between diamagnetic, paramagnetic and ferromagnetic substances.
- to understand why ferromagnetic substances can be magnetized.
- to understand the hysteresis and how the hysteresis loop can be used to determine substances that would make good electromagnets or permanent magnets.

6.1 Introduction

Individual atoms join together to form molecules of liquids and solids. Various forces bind them and give them their properties. Thus, different substances have their own characteristics. Some conduct electricity, others exhibit magnetic properties, some can break easily, others have more elasticity. The use that materials are put to depends on their properties and behaviour.

In this chapter, classification of solids, their mechanical, electrical and magnetic properties will be discussed.

6.2 Classification of Solids

Solids are classified into different types according to the arrangement of the constituent atoms and molecules.

Unlike liquids and gases, solids have definite structure and are dense. They contain a large number of atoms per unit volume, about 10^{28} atoms per cubic meter. The atoms are not scattered around. They are always lined in precise rows and columns.

The bonds (forces) between atoms in solids must explain all these properties and many more — shape of solids, their springiness, their strength, their colour and how they change with temperature.

Sometimes we do not get a true solid but a hard substance, for example, glass. The atoms in glass are arranged more as in a liquid than in a solid.

6.2.1 Crystalline Solids

Most solids are crystalline in nature. Their atoms and molecules are arranged in regular, repeated patterns. A crystal is characterized by the presence of long-range order in its structure. Each atom has strong, equal, cohesive forces binding it. It has an ordered structure.

Some crystalline solids can be held together by covalent bonds. In a diamond crystal, there is an array of carbon atoms, each sharing electron pairs with four other carbon atoms adjacent to it. Diamonds are very hard and have to be heated to temperatures of over 3500°C to break the crystal structure and melt it (Figure 6.1).

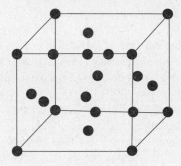

Fig. 6.1: A Diamond Crystal.

The sodium chloride crystal is formed by ionic bonding. One or more electrons from one atom transfer to the other producing a positive ion and a negative ion that attract one another. Ionic bonds are usually strong and so these crystals are strong, hard and have high melting points. The sodium chloride NaCl crystal is cubic in structure as can be visualized by the following figure.

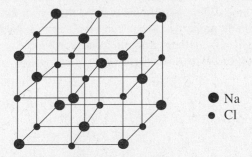

● Na
• Cl

Fig. 6.2: A crystal of NaCl.

Each crystalline solid melts at a sharp melting point only. The basic pattern or unit in a crystal repeats itself throughout the crystal.

Crystals come in all varieties of shapes. An ice crystal is hexagonal while sodium chloride, cesium chloride and copper crystals are basically cubic. Cesium chloride has a chlorine atom at each corner of the cube and a cesium atom at the centre of the cube. Copper is a face-centred cubical crystal. It has copper atoms at each corner of the cube and at the centre of each face of the cube. Crystals have been studied using x-ray techniques.

6.2.2 Amorphous Solids

Some solids lack the definite arrangement of atoms and molecules, which are so clear in crystals. They can be thought of as liquids whose stiffness is due to very high viscosity. Examples of such amorphous solids are pitch, glass and many plastics. The structures of amorphous solid exhibit **short-range order** only. Some substances, for example, boron oxide B_2O_3 can exist in crystalline and amorphous forms. The crystal and amorphous forms have long-range and short-range order respectively. Each boron atom is surrounded by three larger oxygen atoms. In the B_2O_3 crystalline form, a long-range order is present as shown in Figure 6.3.

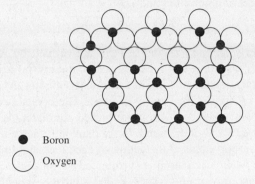

● Boron

○ Oxygen

Fig. 6.3: Crystal form of boron oxide B_2O_3 showing long-range order.

B_2O_3 amorphous forms lack this long-range order. It is a glassy material. It does not have the continuous order and regularity exhibited by the crystalline form (Figure 6.4). It has a disordered structure.

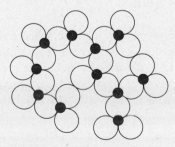

Fig. 6.4: Amorphous form of boron oxide B_2O_3.

This means that some bonds are weaker than others. They can be broken at lower temperatures and the solid gradually softens. As the temperature increases, other bonds break and the **solid melts over a long range of temperatures.** In crystalline solids, all bonds break at the same time and melting occurs suddenly at one temperature only.

6.2.3 Polymers

Polymers are made up of long chain molecules. The long chains combine to form solids. The basic units or monomers combine together at the molecular level to form these long chains. This process is called polymerization. Rubber is a naturally occurring polymer. It has a long chain $(C_5H_6)_n$. Small carbon-hydrogen molecules together with other elements link together into very long chains called macromolecules. These macromolecules have their own special material properties. They have strength, yet they are light weight with low specific gravities.

Synthetic resins bakelite, rayons and plastics have been made artificially. These synthetic polymers are generally both crystalline and amorphous. Polyethylene is about 75 % crystalline. Poly vinyl chloride commonly known as PVC, polytstrene and celluloid are examples of polymers.

Certain formulae of plastics are long lasting and durable as well. Plastics can be fabricated into all sorts of shapes and strengths from television cabinets and compact discs to shoe soles and chappals. Plastic is used more than steel or aluminium and copper.

6.2.4 Crystal Lattice

Crystalline solids have orderly atomic structures. They are formed by repetitive, three dimensional structures called **lattice.** The basic unit that repeats itself in the crystal is called a **unit cell**. In Figure 6.2, one unit cell in the crystal structure of sodium chloride (common salt) is shown. In a NaC1 crystal, each Na^+ ion is surrounded by six Cl^- ions and vice versa. The crystals come in the shape of a cube. The sodium (larger) and chlorine (smaller) atoms are placed alternately at the corners of a cube.

Figure 6.1 is a unit cell of a crystal of diamond. In the diamond, each corner of the cube has a carbon atom. In addition, there are carbon atoms at the centre of each face and four carbon atoms in the cube. It has a tetrahedral structure where each atom is surrounded by four other atoms.

The structure formed by an array of unit cells is called a **lattice**. When crystals break, they fracture along planes parallel to their faces.

Crystals are classified according to their geometrical structures. These include trigonal, tetragonal, hexagonal, triclinic, cubic, rhombic and monoclinic.

6.3 Mechanical Properties of Solids

Solids have been considered as objects having a definite shape and volume. In mechanics, solids and their interactions were studied by assuming they remained un-deformed when external forces were applied on them. Every object in fact is deformed when acted upon by a force. Its size and/or shape may change. It may return to its original shape and size, after the force is removed, or remain permanently distorted.

6.3.1 Deformation in Solids

Solids, when deformed or distorted (squeezed, pressed, stretched, pulled, twisted), return to their original shape and size when the distorting forces are removed. All rigid bodies are to certain extent elastic. This is true for most solids. Such solids are termed 'elastic'. Imagine the forces binding the atoms of the solids as elastic springs (Figure 6.5). We already know that springs return to their original shape when stretched or compressed.

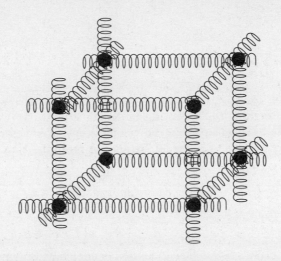

Fig. 6.5: Atoms of a solid bonded by imaginary springs. .

Thus, if an external force is applied (Figure 6.6) to some of these atoms, the 'spring' will compress or expand and when the force is removed, it will come back to its original position.

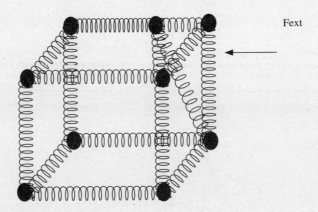

Fext

Fig. 6.6: An external force distorting the molecule.

Different types of distorting forces and stresses may be responsible for these elastic deformations that put a strain on bodies.

6.3.2 Stress and Strain

We have studied in the last section that when different kinds of external forces are applied on solids, they suffer elastic distortion. This distortion can be a change in length, change in volume or a shear may be produced.

The amount of force per unit area that causes these deformations is called **Stress** σ. During the time, the body is subjected to these forces, it experiences stress. The unit for stress in SI system of units is Newton per square metre **N/m² or pascal (Pa).**

$$\text{Stress} \quad = \quad \frac{\text{Force}}{\text{Area}}$$

$$= \quad \frac{F}{A} \qquad\qquad 6.1$$

Strain is a unit deformation. The bodies because of stresses experience changes in length ΔL, shear Δx and in volume ΔV. The change in length per original length strain ε, $\Delta L/L_o$; shear per length $\Delta x/L_o$ (shear strain γ), and change in volume per original volume $\Delta V/V_o$ volumetric strain, are effects of the applied stress. All these are referred to as **Strain. Strain has no units.**

$$\text{Strain} \quad = \quad \frac{\text{extension}}{\text{original length}}$$

$$= \quad \frac{\Delta L}{L_o} \qquad\qquad 6.2$$

This strain can be due to tensile stress related to tension or stretching or due to compressive stress.

$$\text{Shear Strain} \quad = \quad \frac{\text{shear}}{\text{original length}}$$

$$= \quad \frac{\Delta x}{L_o}$$

$$= \quad \tan\theta \qquad\qquad 6.3$$

$$\text{Volumetric Strain} \quad = \quad \frac{\text{change in volume}}{\text{original volume}}$$

$$= \quad \frac{\Delta V}{V_o} \qquad\qquad 6.4$$

Figure 6.7 gives the three types of deformations.

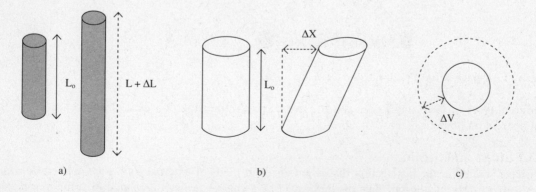

Fig. 6.7: a) Deformation in length, *b)* shear deformation and *c)* volumetric deformation.

Stress and strain are directly proportional to each other:

$$\text{Stress} \quad \alpha \quad \text{Strain}$$
$$= \quad \text{modulus} \times \text{strain}$$

The constant of proportionality modulus is the elastic constant.

6.3.3 Elastic Constants
The three types of distortions are dealt separately and give rise to the elastic constants:
  i. Modulus of elasticity
 ii. Shear Modulus and
iii. Bulk Modulus

6.3.3.1 Elastic Deformation
Strong forces exist between the atoms of a solid. Thus, to distort a solid, a large force would have to be applied. This force would depend on various factors. For rods of the same material, it is found that stretching forces depend on:

i. Amount of stretching
    More force would be needed for more stretching.

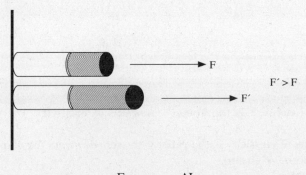

$$F \quad \alpha \quad \Delta L$$

ii. Area of cross-section
    For equal stretching more force would be needed for a rod with more area of cross-section A

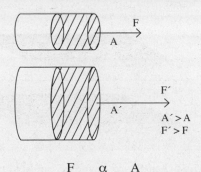

$$F \quad \alpha \quad A$$

iii. Length of rod

A longer rod requires less force to stretch it through the same length compared to a shorter rod.

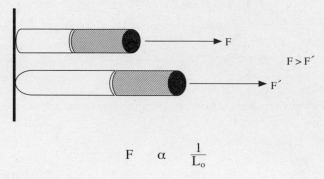

$$F \quad \alpha \quad \frac{1}{L_o}$$

L_o is the original length of the rod.

The stretching force F is thus directly proportional to the change in length ΔL, the area of the rod and inversely to the original length L_o of the rod. It is given by:

$$F \quad \alpha \quad \frac{(\Delta L)\,(A)}{L_o}$$

$$F \quad = \quad Y\frac{\Delta L}{L_o}\,A$$

$$Y \quad \cong \quad \frac{F}{(\Delta L/L_o)\,A}$$

Re-writing
$$= \quad \frac{F/A}{(\Delta L/L_o)} \qquad\qquad 6.4$$

The constant of proportionality **Y is the Young's modulus of elasticity.** The value of Y depends on the type of material.

The Young's modulus of elasticity is the ratio of the tensile stress (or compressive stress) to the tensile strain (or compressive strain).

The SI unit for Y is N/m².

Solids normally have large Young's modulus as large forces are needed to change their length.

6.3.3.2 Shear Deformation

If two strong forces are applied on two opposite faces of a solid in opposite directions, a shear deformation would occur (Figure 6.8).

Fig. 6.8: Shear deformation due to two equal and opposite forces applied to two opposite faces of the cuboid.

It is found that the force is given by

$$F \quad \alpha \quad \frac{(\Delta x)\,(A)}{L_o}$$

$$F \quad = \quad S\frac{\Delta x}{L_o}\,A$$

$$S \quad = \quad \frac{F}{(\Delta x/L_o)\,A}$$

$$= \quad S\frac{\Delta x}{L_o}\,A$$

Re-writing
$$S \quad = \quad \frac{F/A}{(\Delta x/L_o)}$$

Applying equation 6.3 gives
$$S \quad = \quad \frac{F/A}{\tan\theta} \qquad\qquad 6.5$$

S is the Shear Modulus, L_o the separation of surfaces and Δx the displacement of surfaces. **The unit of S is N/m^2.**

Table 6.1 gives Young's and Shear moduli for some important solids.

Table 6.1: Young's Modulus Y and Shear Modulus S for some Solids

Material	Y (N/m^2)	S (N/m^2)
Aluminium	6.9×10^{10}	2.4×10^{10}
Copper	1.1×10^{11}	4.2×10^{10}
Pyrex glass	6.2×10^{10}	–
Steel	2.0×10^{11}	8.1×10^{10}
Tungsten	3.6×10^{11}	1.5×10^{11}

It can be observed that the tensile force of Young's Modulus is parallel to the area A while the shear force is perpendicular to the area. (Please note that area direction is taken perpendicular to the surface.)

6.3.3.3 Bulk Deformation

In bulk deformation, forces are applied on all sides of the body which would distort the body and decrease its volume. The force per unit area is the pressure. Thus, the volume change $\Delta V/V_o$ of a solid is proportional to the change in the applied pressure.

$$\Delta P \quad \alpha \quad \frac{\Delta V}{V_o}$$

$$\Delta P \quad = \quad -B\frac{\Delta V}{V_o}$$

As P = F/A,
$$B \quad = \quad -\frac{F/A}{(\Delta V/V_o)} \qquad\qquad 6.6$$

The negative sign shows that when the pressure increases the volume decreases.

B is the bulk modulus. Its unit are N/m² as well.

Table 6.2 gives the bulk modulus for certain solids and liquids.

Table 6.2: Bulk Modulus B of some common materials

Material	Bulk Modulus B (N/m²)
Aluminium	7.1×10^{10}
Copper	6.7×10^{10}
Pyrex glass	6.1×10^{9}
Steel	1.4×10^{11}
Liquids	
Oil	1.7×10^{9}
Water	2.2×10^{9}

Example: A beam holds a load of 0.8×10^5 N. If the beam is 5 m in length, made of steel with a cross-sectional area 8×10^{-3} m², how much is it compressed? $Y_{steel} = 20 \times 10^{10}$ pa.

$$F = 0.8 \times 10^5 \text{ N}$$
$$A = 8 \times 10^{-3} \text{ m}^2$$
$$Y_{steel} = 20 \times 10^{10} \text{ pa}$$
$$L_o = 5 \text{ m}$$

$$F = Y\left(\frac{\Delta L}{L_o}\right) A$$

$$\Delta L = \frac{(0.8 \times 10^5 \text{N})(5\text{m})}{(8 \times 10^{-3}\text{m}^2)(20 \times 10^{10})}$$

$$= \frac{4 \times 10^5}{160 \times 10^7}$$

$$= 0.025 \times 10^{-2}$$

$$= 2.5 \times 10^{-4} \text{ m}$$

Test yourself: A brass wire with Young's Modulus 9.5×10^{10} Pa is 2 m long and a cross-sectional area of 5 mm². If a weight of 6 kN is hung from the wire, how much does it stretch?

6.3.4 Elastic Limit and Yield Strength

We have already studied that **stress is directly proportional to strain. This is known as Hooke's Law** named after its discoverer Robert Hooke (1635 – 1703).

When the stress is removed, bodies come back to their original size and shape. But if it is stretched too much, they may become permanently distorted. The point at which the body does not return to its original size and shape is called the **elastic limit**. Beyond the elastic limit, distortion occurs.

This can be seen from the following graph (Figure 6.9) between stress and strain.

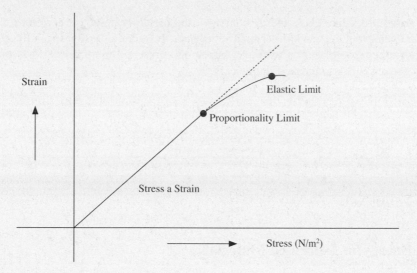

Fig. 6.9: Graph of stress versus strain.

The graph is linear for sometime showing that stress and strain are directly proportional. The slope represents the Young's Modulus, Y.

At the point of the **proportionality limit**, the graph deviates from a straight line to a curve. Although stress is no longer directly proportional to the strain, the body will still return to its original state after the stress is removed. Beyond point of **elastic limit**, the body will no longer come back to its original shape and size when the stress is removed. It will remain distorted. This is termed plasticity.

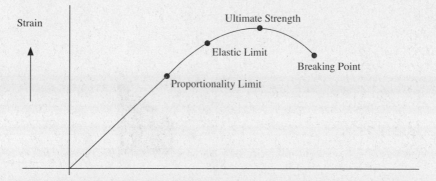

Fig. 6.10: Graph showing the proportionality limit and beyond.

For stresses beyond the elastic limit the solid fractures when the stress reaches the **breaking point.** The stress that a body can bear without breaking is called **ultimate strength** (Figure 6.10).

Materials can be **ductile** or **brittle.**

Ductile materials can stretch beyond their ultimate strength before finally reaching their breaking point. Gold and copper are soft metals. They can be stretched becoming thinner and thinner until eventually they reach the breaking point and break. Other ductile materials are wrought iron and lead.

In **brittle substances** like glass and carbon steel, the breaking point is very close to the ultimate strength. Bone is an excellent example of such substances. It fractures easily, specially in older people, as their bones further lose elasticity with age, becoming more and more brittle. It is the shear stress that normally causes bones to break (Figure 6.11).

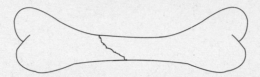

Fig. 6.11: A bone fracture.

6.4 Strain Energy in Deformed Materials

As more stress is applied, strain increases. The relationship between the two quantities is linear. The application of stress and the resulting strain can be compared to a force being applied on a spring which is stretched. To achieve this, work is done which increases the potential energy of the spring. This potential energy is also called **Strain Energy.**

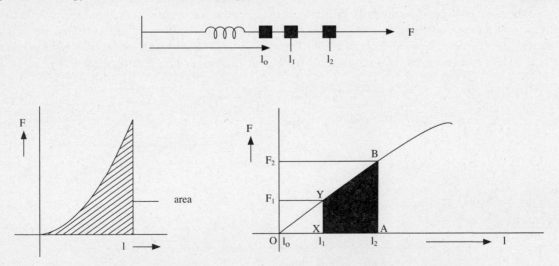

Fig. 6.12: Strain Energy shown graphically.

The area under the curve is the work done and is equal to the strain energy. Assuming that force remains constant in stretching the spring through small changes Δx in length, the strain energy from l_0 to length

l_2, can be computed by adding all the small amounts of work done in stretching l_o to length l_2. From the graph, the energy stored is equal to the area of the triangle OAB.

$$\text{Strain Energy} \quad = \quad \text{area of } \Delta\text{OAB}$$

$$= \quad \tfrac{1}{2} \, l_2 \, F_2$$

Using Young's Modulus

$$F_2 \qquad = \qquad Y\left(\frac{l_2}{l_o}\right) A$$

$$\text{Strain Energy} \quad = \quad Y\left(\frac{l_2{}^2}{2\,l_o}\right) A$$

The Strain Energy is the energy stored in the body due to work done on the body by the applied force in distorting it.

If the body stretches from l_1 to l_2, the strain energy would be the shaded position of the curve, i.e. XYBA.

6.5 Electrical Properties of Solids

Now the electrical property of conduction of electricity will be discussed. There are certain metallic solids like copper and silver which are excellent **conductors** of electric current while others like wood and diamond do not conduct electricity. They are called **insulators**. Between insulators and conductors is a third group, the **semiconductors**, which can conduct under certain conditions.

Insulators have very high resistivities. The resistivity of a diamond is 10^{28} (!) times higher than that of a copper conductor ($\rho_{cu} = 2 \times 10^{-8} \ \Omega.m$). Silicon, a semiconductor, has resistivity of $3 \times 10^{-3} \ \Omega.m$. Its resistivity decreases with the temperature, unlike that of most metallic conductors.

The Bohr theory of free electrons was unable to explain why certain substances are conductors and semiconductors or insulators. The band theory based on quantum physics was able to provide the answers to the electrical properties of materials.

6.5.1 Band Theory

To explain the band theory, let us consider a copper atom. A copper atom has 29 electrons which are arranged in 6 filled subshells (Figure 6.13). The 4th subshell has only one electron. Each subshell can hold two electrons.

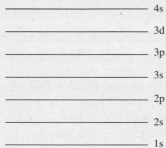

Fig. 6.13: Six subshells of copper atom.

If two or more copper atoms are brought closer together, their wave functions will overlap. The two atoms are no more separate entities but a **two-atom system** of 58 electrons. Each of these 58 electrons must occupy a different quantum state. Each energy level of the lone copper atom will now split into two levels or states for the two-atom system (Figure 6.14).

Fig. 6.14: Splitting of each energy level into two levels for a two-atom system.

In a crystalline solid, there are many atoms arranged in repeating structures called a lattice. **If there are n atoms in a lattice, each energy level would split up into n sub-levels.** These sub-levels are very close together and appear as **energy bands.**

   The uppermost band is the conduction band. The band below is the valence band. The energy bands are separated by energy gaps where no electron can exist (Figure 6.15).

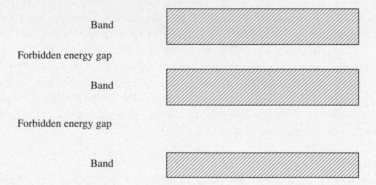

Fig. 6.15: Energy bands and forbidden energy states (gaps).

6.5.2 Insulators

When a potential is applied across a solid and there is no current, the solid is an insulator. For a current to flow, the electrons must acquire kinetic energy. The valence electrons in insulators are tightly bound to their individual atoms and so cannot conduct electricity.

   According to the band theory, the upper band which is the valence band containing electrons is full. To gain energy, electrons must jump to higher orbits. These are already filled, so the electrons are unable to acquire energy. There is an unfilled band above the filled valence band but to jump the huge gap to the next band (conduction band), very high energies are needed (Figure 6.16). In insulators, these energy gaps are in the range of 3 – 5 eV and are so insurmountable.

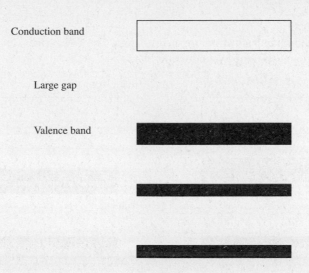

Fig. 6.16: The gap between the top filled band and the unfilled band is too wide.

6.5.3 Conductors

In. contrast to the energy bands of insulators, the upper most valence band in metals is half-filled. The conduction and valence bands in metals are very close to each other. They might even overlap. Thus even with small energies (1 eV), electrons can jump from the valence and into the conduction band (Figure 6.17).

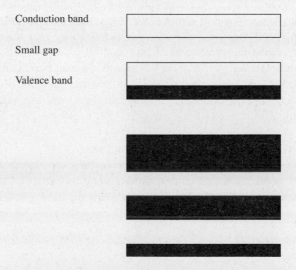

Fig. 6.17: Half-filled valence band, conduction and valence band close to each other, sometimes even overlapping.

In metals, the valence electrons are not tightly bound to the atoms. They are free to move and conduct electricity. They are called **free electrons**.

6.5.4 Semiconductors

Semiconductors have the same band structure as insulators. The difference is in the energy gap between the topmost filled band and the empty conduction band.

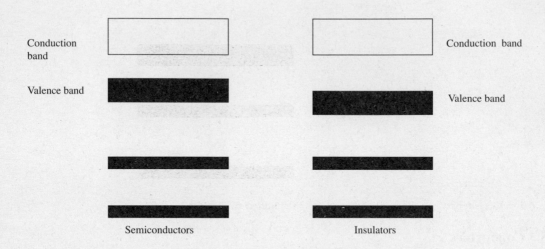

Fig. 6.18: Energy bands in semiconductors and insulators.

The forbidden energy gap is much less in semiconductors as compared to insulators as can be seen in Figure 6.18.

At temperatures below zero (0°C), the semiconductor is like an insulator. The valence band is completely filled. At room temperature, in semiconductors, some electrons gain energies and move into the conduction band. The semiconductor starts to conduct slightly. The electrons that move into the conduction band leave a hole in the valence band.

For a silicon semiconductor, the gap is 1.1 eV. Thus, if the electrons can acquire enough energy, they can jump the gap to the next band creating holes in the valence band. Both holes (valence band) and electrons in conduction band can conduct. The current through the semiconductor is the sum of the electron and hole currents.

6.5.5 Intrinsic and Extrinsic (Doped) Semiconductors

Intrinsic semiconductors are naturally occurring substances that are absolutely free from impurities. These include crystals of germanium and silicon. Both belong to Group IV (Figure 6.19). They have four electrons in the outermost shell. All semiconductor elements have a valency ± 4.

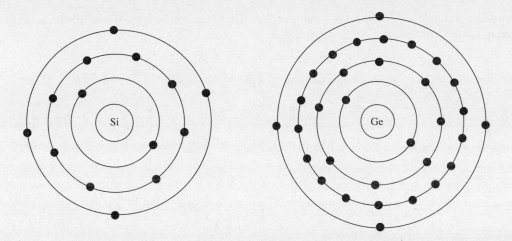

Fig. 6.19: The electronic configuration of atoms of silicon and germanium.

As you are aware that for stability there should be eight electrons in the outermost shell. The atoms of silicon and germanium share electrons with their four immediate neighbours to achieve a configuration of eight electrons in the outermost shell. This is done by covalent bonds (Figure 6.20).

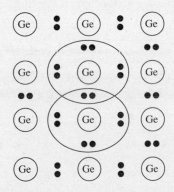

Fig. 6.20: Each germanium atom shares its four valence electrons with four other neighbouring atoms.

The covalent bonds are responsible for the regular pattern that characterizes crystals.

Silicon was discovered in 1823 and germanium in 1886. These two elements are widely used in semiconductor devices. Although germanium has a much lower resistance than silicon, it is silicon that is used more often in semiconductors.

Extrinsic semiconductors are produced by doping the pure crystals with impurities. Impurities are added to the pure crystals by heating the semiconductor crystal with vapours of the impure elements. When the impurity has been added to the crystal, it is called **extrinsic semiconductor or doped semiconductor.**

Donor or n-type are extrinsic semiconductors in which the impurity added is a Pentavalent element. Four of the electrons in the outermost shell of the pentavalent atom form covalent bonds with silicon electrons to complete the shell. The fifth electron is extra (Figure 6.21). It is like a free electron. It is free to move and will conduct current when a potential difference is applied. Pentavalent elements used in doping are arsenic, antimony and phosphorus.

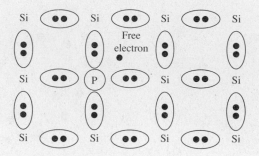

Fig. 6.21: A pentavalent impurity results in free electron.

These substances are called **N-type** as they have free negative charges. They are also called **Donor** because they can donate electrons.

P-type or positive type semiconductors can be produced by adding a trivalent impurity like aluminium, boron, gallium or indium. Covalent bonds are again formed between the valence electrons of silicon and aluminium. But there will be one electron short as aluminium has only three electrons in its valence shell. The outermost shells will have seven electrons. There will be a **hole** charge (Figure 6.22).

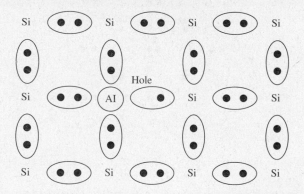

Fig. 6.22: A hole due to trivalent impurity is produced.

In the doped crystals, there will be many such hole charges. They are like positive charges and are capable of moving easily to conduct a current.

These semiconductors are also known as **Acceptor type** substance as a hole can accept a neighbouring electron.

This is shown in the following figure. A hole charge is shown adjacent to filled covalent bonds.

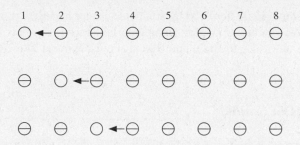

Fig. 6.23: Motion of a hole charge.

The electron in 2 moves to position 1.The hole has moved to 2 and the bond at 1 is now filled. Similarly, if the electron in 3 moves to position 2, the hole has moved to position 3. Since the holes are like positive charges, this motion of a hole from position 1 to position 3 constitutes a current. This will occur when a potential difference is applied across position 1 and 3 with the position 1 at positive potential.

6.6 Superconductors

As we know that even the best conductors offer resistance to the flow of current. The amazing fact is that the resistivity of some of these conductors becomes 4×10^{-25} Ω.m (!) at approximately zero temperature. For example, aluminium has no resistance at $1.18°K$ and tin at $3.72°K$.

Superconductors offer **no resistance** to the flow of charge through them. If a current is set up in a superconducting circuit, the current will last forever and will require no battery or power supply to maintain it.

Superconductivity was discovered in the beginning of the 20th century (1911) by a Dutch physicist, Onnes. While experimenting with mercury, he found its resistivity became zero. As the temperature of the mercury is reduced, its resistivity decreases until at the critical temperature of 4.2 K, it suddenly becomes zero. This becomes a drawback to the development of superconductivity because to achieve and maintain such low temperature is very costly.

It is only recently in 1986 with the discovery of ceramic materials which exhibited superconductivity at much higher temperatures that the superconductivity technology developed. The ceramic superconductors are made of copper oxides mixed with elements like lanthanum and yttrium. These ceramic superconductors are also cheaper to produce. The materials have interesting properties. At temperatures as high as 125 K, they are superconducting, but at temperatures above this temperature, they are insulators. Lian Lee experimenting in Cambridge discovered that yttrium barium copper oxide exhibits at even higher temperature of 163 K. Materials that exhibit superconductivity at temperatures above the boiling point of nitrogen (77 K) are called high temperature superconductors. This is important as cooling can be carried out with liquid nitrogen instead of liquid helium which is more expensive to use.

Metals like silver and copper are good conductors but can never be made into superconductors.

Research is being carried out to find materials which are superconducting at room temperature.

Some important technological applications of superconductivity are cheaper transmission of electrical power, faster computers and levitation of trains.

Superconductors **expel magnetic fields.** Magnetic fields cannot penetrate the superconductors which repel magnets. In Japan, a prototype maglev (**mag**netic **lev**itation) train has been constructed using superconducting magnets to levitate the train upto 10 cm above the track. These trains move with speeds of 300 mph.

6.7 Magnetic Effects of Solids

It has already been pointed out that magnetism, that is magnetic fields and forces, are a result of charges in motion. A bar magnet has a magnetic field. It attracts pieces of iron. We do not see any charges in motion or a current in the bar magnet. Then how does the bar magnet become a magnet?

If the magnetic fields due to a single loop coil and a small magnet and that between a solenoid coil and a bar magnet are compared, they are very similar (Figure 6.24).

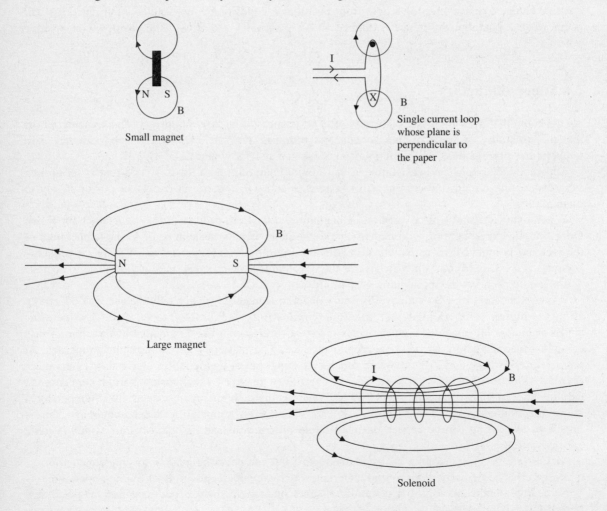

Fig. 6.24: The magnetic fields due to a small magnet and a single current loop (top). Magnetic field due to a bar magnet and solenoid (bottom).

Ampere suggested that the magnetic field in a bar magnet must also be due to moving charges but where were the moving charges in a piece of metal?

Later with the discovery of the atomic structure, it was found that magnetic fields were produced due to the motions of the electrons and nucleus of an atom. The orbital motion of electrons as well as their spins produce magnetic fields. The nucleus with its positive charge too is spinning and produces a very small magnetic field. The electron also has an intrinsic magnetic dipole moment that is a property of the electron. It is independent of its motion. The magnetic fields due to the individual atoms are both due to the motions of the electrons and the nucleus and spins. They can add up and give a resultant field or cancel out and there is then no net magnetic field. The orbital magnetic field is greater than the spin magnetic field for electrons in atoms.

If there is a resultant field, the atom behaves as a tiny magnet and is known as a **magnetic dipole.**

Nonmagnetic substance can be divided into 2 categories — **Diamagnetic and Paramagnetic substances.**

6.7.1 Diamagnetic materials

Diamagnetic materials are those substances whose atoms do not have a resultant magnetic field. These substances like copper and gold do not show any magnetic properties. They are called diamagnetic.

6.7.2 Paramagnetic substances

Paramagnetic substances are those substances whose atoms are dipoles although the substance does not behave as a magnet. This is because the dipoles are randomly arranged so that there is no overall net magnetic field, e.g. aluminium, platinum and potassium.

6.7.3 Ferromagnetic substances

Ferromagnetic substances have small regions or domains where thousands of dipoles align in such a way that their magnetic fields are in the same direction. Different domains have such alignments. Each domain is like a small magnet and has 10^{12} to 10^{16} dipoles (atoms). When these substances are placed in an external field, the domains align in the direction of the external field and the substance is magnetized and behaves as a magnet (Figure 6.25).

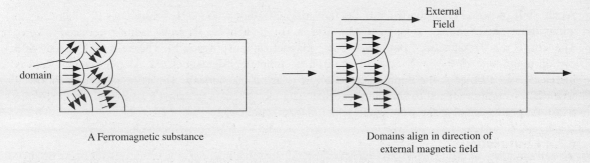

Fig. 6.25: Ferromagnetic substances showing domains. When placed in a strong magnetic field, the domains align to magnetize the material.

Once magnetized, the ferromagnetic substance may not lose its magnetization when the external field is removed. To align the domains, energy is expended in overcoming friction. If there is more friction that has been overcome, then the substance shall remain magnetized. It will become a permanent magnet. If there was little friction, then the domains will easily return to their original position and the substance would be demagnetized. Ferromagnetic substances are iron, nickel and cobalt.

Demagnetization can be achieved by heating. The thermal energy makes the atoms vibrate faster and lose their alignments. Iron at 750°C becomes paramagnetic. The temperature at which substances lose their magnetization varies in different substances and is known as the **Curie temperature.**

6.7.4 Hysteresis

A ferromagnetic substance is magnetized by placing it in an external magnetic field, which is gradually increased to a maximum. The demagnetization process when the field is decreased does not follow the same path. This non-retracability is called **hysteresis.**

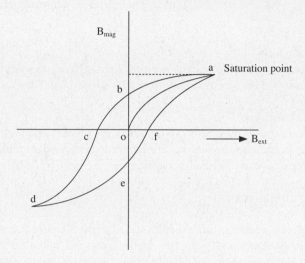

Fig. 6.26: Hysteresis loop.

At the start (position O) when $B_{ext} = 0$, the ferromagnetic shows no magnetization. As B_{ext} increases to a maximum, the iron piece is magnetized (position a). The field B_{ext} is now gradually reduced to zero. The iron is not demagnetized but retains some magnetization (position b). The field is reversed. At a certain magnitude of the reversed field, the iron is completely demagnetized. As the external field is increased still further in the opposite direction, it is again magnetized. The same pattern is followed when B_{ext} is again reduced to zero and reversed. The loop abcdefa is called a **hysteresis loop** (Figure 6.26). At points b and e, the iron retains some magnetization.

6.7.4.1 Saturation

As the magnetizing force/field is increased, the ferromagnetic substance starts getting magnetized. As all the domains align, magnetization is maximum and even if the magnetizing force is further increased, the substance will not become more magnetized. The material is **saturated.**

6.7.4.2 Retentivity

When the magnetizing force/field is decreased, the ferromagnetic substance starts to lose its magnetization but more slowly. When the magnetizing field becomes zero, the material is still partially magnetized. The ferromagnetic substance retains some magnetization. This value of magnetization ob is termed as **Retentivity.** It is also called **residual magnetism.**

6.7.4.3 Coercive Force

To further demagnetize the magnetized substance, the magnetizing field is reversed. The magnetization decreases until at a certain value of this force field oc, the material is totally demagnetized. The value of the force field is known as **coercive force.**

Magnetic materials where the domains remain aligned even after the magnetizing field is removed are called **hard magnetic substances.** These materials would make good permanent magnets. In **soft magnetic substances**, the domains return easily to their random states and the material is easily demagnetized. They would thus be good for producing electromagnets.

6.7.4.4 Hysteresis Curve Area

The area of the hysteresis curve also gives a clue as to whether a material can be used as a permanent magnet or electromagnet. The area represents the amount of energy that has been used up in overcoming friction when the domains were magnetized and demagnetized. More energy used up or dissipated per cycle would mean the material is more difficult to demagnetize. Thus, a material whose hysteresis loop has more area will be more suitable for a permanent magnet (Figure 6.27). Steel is such a material.

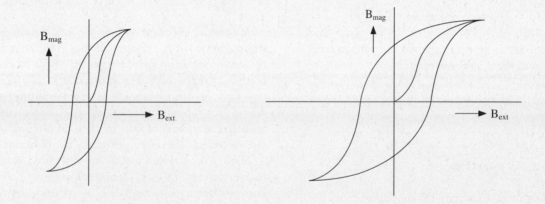

Fig. 6.27: Substances with hysteresis loops of small area are good electromagnets and substances with larger area loops are good permanent magnets.

A hysteresis loop with small area would make a good electromagnet. Soft iron is used in electromagnets.

Summary

Solids are classified as crystalline, amorphous and polymeric solids.

Deformation in solids (strains) results from stress, which is force exerted per unit area.

$$\text{Stress } (\sigma) \; = \; \frac{\text{Force}}{\text{Area}}$$

Strain in one dimension is tensile strain or can be compressive strain. It is given by

$$\text{Strain} \; = \; \frac{\text{Change in length}}{\text{Original length}}$$

$$= \; \frac{\Delta L}{L_o}$$

When stress changes volume, volumetric strain results

$$\text{Volumetric strain} \; = \; \frac{\Delta V}{V_o}$$

When shear stress is applied to opposite faces of a cube, shear strain is produced

$$\text{Shear strain} \; = \; \frac{\Delta x}{L_o}$$

$$= \; \tan \theta$$

Elastic constants are

Modulus of Elasticity (Young's)

$$Y \; = \; \frac{F/A}{\Delta L/L_o}$$

Bulk constant

$$B \; = \; \frac{F/A}{\Delta V/V_o}$$

and
Shear constant

$$S \; = \; \frac{F/A}{\tan \theta}$$

When a body is stretched, it returns to its original state before the elastic limit. Beyond the elastic limit, deformation occurs. The ultimate tensile strength is the maximum strain a material can withstand. Beyond this, the material can break. Materials are classified as ductile and brittle.

In deformed materials, strain energy can be found by

$$\text{Work Done} \; = \; \frac{1}{2} \frac{Y \, A \, \ell_1^2}{L}$$

The band theory is able to explain the conduction of current in insulators, semiconductors and conductors. The forbidden gap between valence and conduction bands is larger between insulators than in semiconductors. In conductors, the two bands overlap or are close to each other.

Intrinsic semiconductors are pure substances while extrinsic have impurities (trivalent/pentavalent) in the ratio $1:10^6$.

Superconductors are substances whose resistivity becomes zero below the critical temperature T_c.

Magnetic substances can be classified as **diamagnetic** whose atoms do not exhibit any magnetic properties; **paramagnetic** whose atoms behave as dipoles but the substance as a whole does not exhibit magnetism; **ferromagnetic** whose atoms behave as dipoles. The dipoles are arranged in groups of 10^{12} to 10^{18}, all aligned in one direction. These groups are called **domains**.

High temperatures affect the domains. At a temperature of 750°C (Curie temperature), iron converts from ferromagnetic to paramagnetic. When ferromagnetic substances are placed in a magnetic field, they are magnetized as the magnetizing current increases. As the current is brought to zero, some magnetism is retained by the substance. It is termed as **retentivity**. Current is reversed till the substance is demagnetized. This is the **coercive strength**. The **hysteresis loop** measures the energy needed to magnetize and demagnetize a substance. Greater area of the hysteresis loop would make a good permanent magnet.

Questions

1. Distinguish between crystalline and amorphous substances.
2. How are crystals classified?
3. What are the units of strain?
4. What is the difference between tensile strain and compressional strain?
5. Distinguish between Young's Modulus and Bulk Modulus?
6. How do materials react to stress before the elastic limit?
7. What is the difference between insulators and conductors according to the band theory?
8. What is the ratio of the doping atoms in semiconductors?
9. Distinguish between extrinsic and intrinsic materials.
10. Why certain doped substances are called donor type substances?
11. What are bands?
12. What is critical temperature?
13. What are the critical temperatures of mercury, lead, aluminium?
14. What are high temperature superconductors?
15. What is a magnetic dipole?
16. Differentiate between paramagnetic and diamagnetic substances?
17. Why is iron a poor permanent magnet?
18. Discuss the significance of the Curie temperature?
19. What is retained in magnetic retentivity?
20. Why is the hysteresis loop important in determining which material can be used for permanent magnets?

Problems

1. The molecules or ions in a crystalline solids are:
   a) stationary   b) not stationary
   b) in random motion   d) none of these
2. Amorphous substances melt
   a) at a particular temperature for that substance
   b) melt at no particular temperature
   c) melt slowly in a range of temperatures
   d) none of these
3. The unit of stress is
   a) N/m   b) Nm
   c) N/m^2 d) No unit
4. Tensile stress changes
   a) Volume   b) Length
   c) Area   d) Shape
5. Permanent deformation produced in materials is termed as
   a) elasticity   b) proportionality
   b) plasticity   d) none of these
6. A steel wire of diameter 1 mm can support tension of 0.2 kN. A cable supportig a tension of 20 kN should have a diameter of what order of magnitude.
7. The heels on a pair of women's shoes have a radius of 0.5 cm at the bottom. If 35% of the weight of a woman weighing 450 N is supported by each heel, find the stress on each heel.
8. A hair breaks under a tension of 1.2 kN. What is the diameter of the hair? The tensile strength is 2×10^8 Pa.
9. For safety in climbing, a mountaineer uses a polyester rope that is 55 m long and 1 cm in diameter. When supporting a 80 kg climber, the rope elongates 2.0 m. Find its Young's Modulus.
10. A man whose weight is 1.0 kN is standing upright. By how much is his femur shortened compared to when he is lying down? Assume compressive force on each femur is about half of his weight. The average cross-sectional area of the femur is 7.5 cm^2 and length of femur when lying down is 42 cm. (Young's Modulus for femur = 9.4×10^9 Pa)
11. Bone has a Young's Modulus of 18×10^9 Pa. Under compression it can withstand a stress of about 160×10^6 Pa before breaking. Assuming that a femur is 0.45 m long,

calculate the compression this bone can withstand before breaking.

12. The upper surface of a cube of soft rubber 6 cm side is displaced 0.5 cm by a tangential force. If the shear modulus of the soft plastic is 1000 Pa, what is the magnitude of the tangential force?

13. An elevator of mass 3000 kg moves upwards with an acceleration of 0.5 m/s^2. The steel cable of the elevator has a maximum safe working stress of 1.5×10^8 Pa. What is the cross-sectional area of the steel cable? (g = 9.8 m/s^2)

14. What is the maximum load that could be suspended from a copper wire of length 1 m and radius 1mm without deforming. Copper has an elastic limit of 2×10^8 Pa and a tensile strength of 4×10^8 Pa.

15. A solid brass sphere of volume 0.5 m^3 is dropped in the ocean to a depth of 2000 m. The pressure increases by 2×10^7 Pa. What is the change in volume of the sphere?

7 Electronics

OBJECTIVES

After studying this chapter, you should be able:
- to understand the working and conduction process of the p-n junction.
- to understand the application of p-n junction in rectification.
- to understand the npn and pnp transistors, forward and reverse biasing and conduction in transistors.
- to understand the applications of transistors in full-wave rectification, amplification and as a switch.
- to understand the idea of gain in transistors.
- to understand the operational amplifiers and their characteristics.
- to understand the operational amplifier as an inverter, non-inverter, comparator and switch.
- to understand the basis of digital electronics.
- to understand the logic gates – AND, OR, NOT and others NAND, NOR, XOR, XNOR.
- to understand the application of logic gates in a control system.

7.1 Introduction

With the discovery of the electromagnetic waves and radio communication, the foundation was laid for modern electronics. The most important invention in electronics was the transistor in 1948. It took the place of unwieldy vacuum tubes, and soon the world saw integrated circuits (ICs). Smaller and still smaller sized circuits were developed. Today, chips are integral to all computer circuitry. They are responsible for more sophisticated and miniature circuits, and are also inexpensive and reliable. In this chapter, the p-n junction, operational amplifiers and gates shall be discussed.

7.2 p-n Junction Revisited

Pure silicon at room temperature allows a very small amount of current to flow when an electric field is applied across it. Silicon or germanium can be made more conducting if impurities (about 1 in a million) are added. The p-n junction is made by adding n-type impurity in one half of a crystal of silicon (or germanium) and doping the other half with p-type impurity. The n-type half has electrons that are free to move about although the crystal as a whole remains electrically neutral. The p-type has holes (which behave like positive charges) into which a neighbouring electron can move.

In the semiconductor diode as the doped p-n junction is called, conductivity is more than in pure silicon and germanium. In Figure 7.1, the bold central line which is actually very thin ($\sim 10^{-3}$mm) represents the junction between the p-type and n-type halves. The p-n junction is the basis of all semiconductor devices and the microelectronics industry.

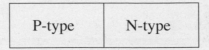

Fig. 7.1: A p-n junction, n-type impurity in half the silicon crystal and p-type in the other half.

Although both p-type and n-type substances are neutral, the n-type has free electrons and the p-type holes. Conduction is motion of holes/electrons across the p-n junction.

As soon as the crystal with impurities is formed, electrons from the n-type will cross the p-n junction and collect across the barrier on the p-type side. Holes will also move across and collect on the n-type side. Across the junction, charges accumulate. Due to this charge build up, a potential difference V_o is generated across the junction. A small region known as the **depletion zone** between this charge build-up of width approximately 10^{-4} cm is produced where there is no movement of charges (Figure 7.2).

Thus, conduction is restricted. An electron trying to cross the junction is repelled by the negative charge on the p-side and a hole trying to move in the other direction will also be repelled by holes on the n-side of the barrier. **The barrier potential difference for silicon is 0.3 V and for germanium it is 0.7 V.**

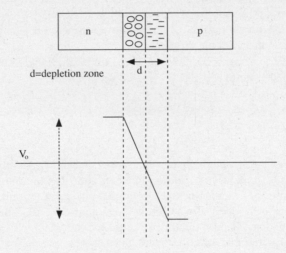

Fig. 7.2: Distribution of charges and holes across the p-n junction and the barrier potential V_o.

In the semiconductor diode, the p-type is the anode and the n-type the cathode. The symbol for the p-n junction is

7.2.1 Forward Bias
The p-n junction can be made conducting if the barrier potential is reduced. This is achieved by connecting the p-type end of the semiconductor to the positive terminal and the n-type to the negative terminal of the battery. Figures 7.3 and 7.4 show how in forward biasing the potential barrier V_o as well as the depletion zone d are reduced and a current starts to flow.

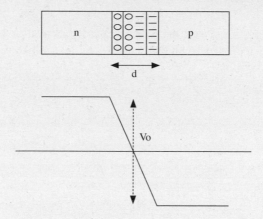

Fig. 7.3: Potential barrier V_o and depletion zone d in a p-n junction.

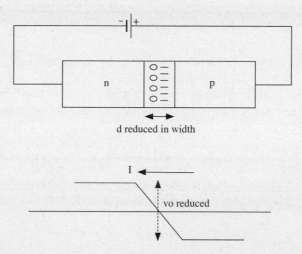

Fig. 7.4: Forward biasing: The depletion zone and barrier potential are reduced. The p-n junction starts to conduct.

Electrons are attracted to the positive of the battery and holes to the negative. This reduces the barrier potential; diffusion takes place and the p-n junction is in **forward-bias connection**. The p-n junction is said to be conducting. The depletion zone is reduced as well.

A net forward current I_F of a few milliamperes flows in the circuit. As the biasing voltage is increased, the current I_F also increases (Figure 7.5). The forward bias resistance R_F can be computed from the graph. It is equal to

$$\text{Resistance}_{forward} \quad = \quad R_F$$

$$= \quad \frac{\Delta V_F}{\Delta I_F} \qquad\qquad 7.1$$

This is the resistance offered when the p-n junction is conducting.

7.2.2 Reverse Bias

If the battery is connected the other way, the potential barrier as well as the depletion zone increases and the diffusion current decreases. A small drift current remains. When the potential is increased, the small drift current remains constant.

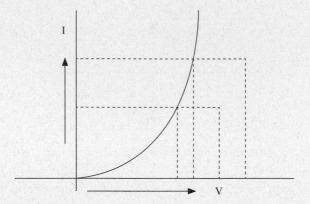

Fig. 7.5: p-n junction forward current increases as forward biasing voltage increases.

If the reverse biasing (potential) is raised to a high value, the covalent bonds between the atoms of the crystal break and a high current is recorded. The crystal structure has been destroyed (Figure 7.6).

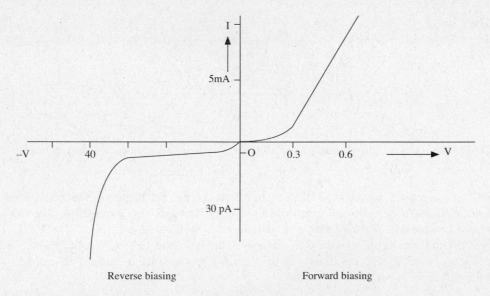

Fig. 7.6: Forward biasing: Current increases as voltage is increased.
        Reverse biasing: Current hardly changes as voltage is increased.

The p-n junction only conducts in the forward bias mode. This is an important property of the p-n junction. It only allows the current to flow in one direction when it is forward biased.

7.2.3 Variation of current with voltage

In the forward bias mode, as the potential difference across the p-n junction V is increased, the current I increases as well. When reverse biased, the current is very small and even for high voltages its value remains small. At very high voltages, the current increases dramatically as the crystal structure breaks down. The covalent bonds are destroyed and electrons are available resulting in a high current (Figure 7.6).

The circuits for forward and reverse biasing are given in Figure 7.7.

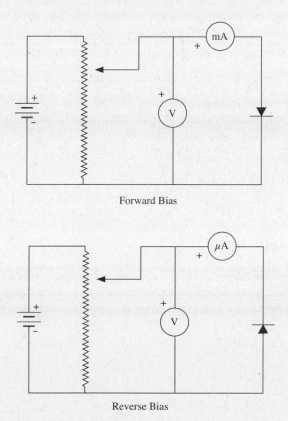

Forward Bias

Reverse Bias

Fig. 7.7: Top: Circuit diagram for forward biasing.
　　　　Bottom: Circuit diagram for reverse biasing.

When the semiconductor diode is connected in forward bias mode (positive of battery to anode and negative of battery to cathode), it conducts and a current flows. The diode behaves as a switch. Resistance is low. It is on. When connected in reverse bias, no current flows. It is as if the diode is switched off.

7.3 Rectification

The conversion from AC to DC is called rectification.
The p-n junction is very useful in rectification because of its property of only allowing the current to flow in one direction when it is forward biased.

7.3.1 Half-Wave Rectification
A p-n junction is connected to an AC power supply through a transformer. The output V_{out} is across the load resistance R_L. Each AC cycle has a positive half cycle and a negative half cycle. The circuit for half-wave rectification is shown in Figure 7.8a.

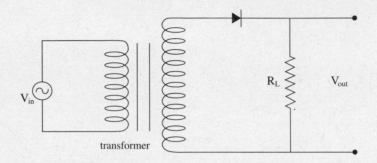

Fig. 7.8a: Circuit for half-wave rectification.

In the first half cycle, the p-type (anode) of the semiconductor diode is connected to the positive half cycle. The diode conducts and an output cycle is obtained. In the second negative half cycle, the semiconductor diode will not conduct and so there is no output during this cycle.

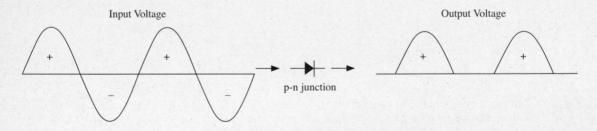

Fig. 7.8b: Half-wave rectification. Only the positive half cycles are rectified.

Thus, only the positive half cycles appear in the output during which the diode is conducting (Figure 7.8b). A capacitor placed across the load resistance can reduce the fluctuating AC element in the output.

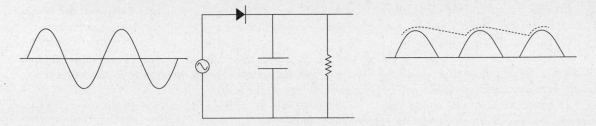

Fig. 7.9: Input and output waveforms for half-wave rectification.

The capacitor charges up as the voltage rises and in the second quarter cycle, the capacitor discharges. It prevents the voltage from becoming zero between the positive half cycles. The dashed line in Figure 7.9 is the output from the capacitor. The capacitor circuit is called a filter because it filters out the AC element.

7.3.2 Full-Wave Rectification
A full-wave rectifier rectifies the complete AC cycle, thus preventing the loss of half of the cycles that is incurred in half-wave rectification. The full-wave rectifier connects four semiconductor diodes in a bridge circuit as shown in Figure 7.10.

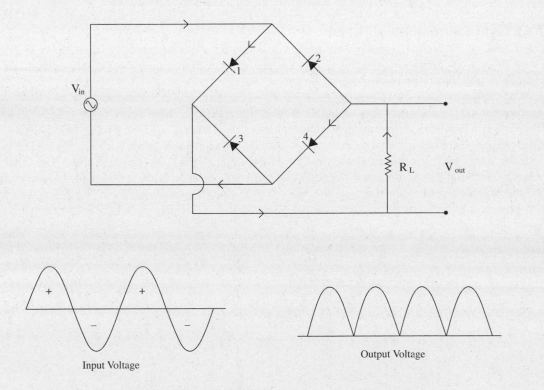

Fig. 7.10: Full-wave rectifier bridge circuit. Also shown are input and output waveforms.

In the first positive half cycle, diodes 1 and 4 will be conducting. Current will flow from the power supply through diode 1 and R_L, through diode 4 and back to the power supply. The output across the load R_L will also be positive.

For the next negative half cycle, diodes 3 and 2 will be conducting. Current will flow through diode 3, R_L and through diode 2 back to the power supply. **Its direction through R_L is the same.** Thus, the negative input half cycle will become positive across the load.

A capacitor across R_L would smooth the fluctuation in the voltage to some extent (Figure 7.11). The capacitor circuit filters out the AC fluctuating voltage.

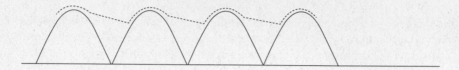

Fig. 7.11: The dotted line shows the output from the capacitor.

7.4 Special p-n junctions

Special p-n junctions are fabricated to allow different functions and applications.

7.4.1 Light-Emitting Diode (LED)
Light-emitting diodes (LED) are mainly used as indicator lights on electronic equipment. We see them everywhere from digits on calculators, watches and clocks to remote control devices. The light in practically all cases comes from LEDs.

When impurities are added to pure semiconductors like silicon and germanium, electron-hole pairs are formed. The formation of these requires energy. When they combine again, this energy is released (Figure 7.12). In silicon and germanium semiconductors, these energies are very small, below the visible light range and are absorbed as thermal energy of the vibrating lattice atoms.

If these energies have a higher value and fall in the visible light range, they are emitted as different colours. To obtain these colour emissions, special semiconductors have to be designed. They are based on gallium doped with arsenic and phosphorus. A common compound used is gallium arsenide phosphate GaAsP.

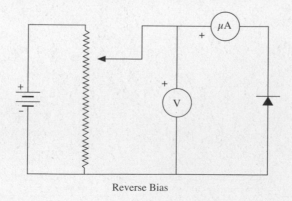

Reverse Bias

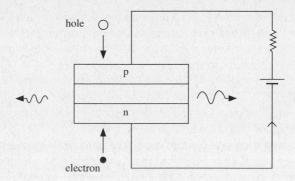

Fig. 7.12: LED: Electrons and holes combine releasing light energy.

An external current supplies the energy necessary to excite electrons to the conduction band. When electrons fall back they combine with holes emitting photons.
An arrangement of seven LEDs can display digits from 1 to 9 (Figure 7.13).

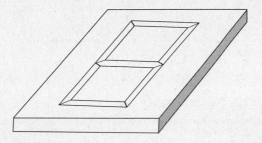

Fig. 7.13: Digit display with 7 LEDs.

7.4.2 Photo Diode

If light is irradiated on a suitable p-n junction, a current can be produced in the circuit. The photo diode works on this principle.

Infrared light is sent from your remote control device (LED) to the receiver (photo diode) on the television controls. These particular light pulses trigger the photo diode which sends a current to put the television on or change the channel. Figure 7.14 shows the circuit for the operation of a photo diode.

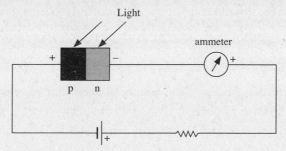

Fig. 7.14: A photo diode connected in reverse bias.

The photo diode is connected in reverse bias. When it is not exposed to light, the micro-ammeter does not register a current. As the light intensity increases, the reverse current also increases. The photo diode is very sensitive. It can detect light very fast in 10^{-9} seconds. It has numerous applications from logic gates and fast switching systems to detection of radiation.

7.4.3 Photovoltaic Cell

Direct conversion of thermal, mechanical and chemical energy to electric energy is carried out by thermoelectric, piezoelectric and chemical actions respectively.

Radiant energy can be converted into electric energy by photovoltaic effect.

When light falls on a p-n junction, an emf is produced. A semiconductor diode is designed to absorb incident light which appears as a potential difference across the junction. The potential difference generates an electric current in the load resistance connected across the junction.

The solar cell is a photovoltaic converter. A thin layer of p-type over a thinner (0.5 μm) n-type layer silicon crystal is formed. The external circuit is a load resistance R_L across the p-n termini (Figure 7.15).

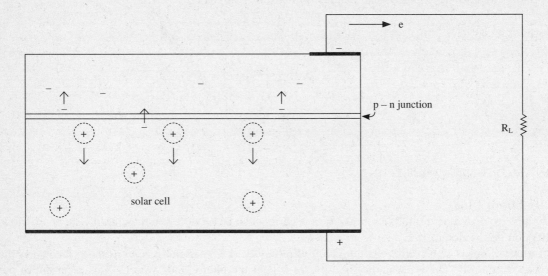

Fig. 7.15: A Photovoltaic Cell.

Light photons pass through the thin n-type layer and are absorbed by the crystal lattice atoms. Electrons gain energy and are ejected, leaving the holes behind. The photons thus produce electron-hole pairs. The electric field of the p-n junction makes the electrons move towards the n-layer and the holes towards the p-layer. Electrons so move through the external circuit (R_L) from the n-layer to the p-layer to fill up the holes there.

The photovoltaic cell acts as a source of emf and the current in the external circuit.

The photovoltaic cell produces small currents of a few milliamperes and voltages of 0.6 V in silicon p-n junctions.

7.5 Transistors

The transistor has been responsible for the billion dollar industry with products ranging from pocket radios, calculators to highly sophisticated computer systems. Like the p-n junction, the transistor is also a device formed by adding impurities in a silicon or germanium crystal. The crystal is doped in such a way that two p-n junction are placed back-to-back so that either the n-type substance is between p-type regions or a p-type region is in the center.

If p-type region is surrounded by n-type, it is a n-p-n transistor and if n-type is in the centre, it is a p-n-p transistor (Figure 7.16).

p–n–p transistor   n–p–n transistor

Fig. 7.16: p-n-p and n-p-n transistors.

The transistor has three terminals. It has two circuits. The currents in the two circuits are linked so that the current in one circuit affects the current in the other circuit. The transistor can amplify currents, voltages and can also act as a switch.

The first region of the transistor is called **emitter,** the middle **base** and the third region the **collector**. The symbol for the p-n-p and n-p-n transistors is shown in Figure 7.17.

The arrow of the emitter gives the direction of the conventional current.

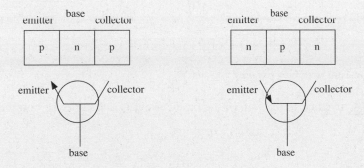

Fig. 7.17: Transistors and their symbols.

The emitter is heavily doped providing more electrons. Usually the base is thin about 10^{-3} mm and the collector region slightly larger than the emitter region.

7.5.1 Current flow in transistors

The emitter-base junction is forward biased (V_{EE}) while the base-collector is reverse biased (V_{CC}) for transistors to operate. The voltage V_{CC} is much larger than V_{EE}.

When connected as shown in Figure 7.18, the electrons cross the emitter-base junction. As the base region is very thin and the base-collector is reversed biased, most of the electrons are attracted to the positive of the battery V_{CC}. They accelerate across the base-collector junction and enter the collector region. About 98% of the electrons move into the collector region. Very few form the small base current I_B.

Although the base current is a very small current, small changes in it significantly alter the collector current.

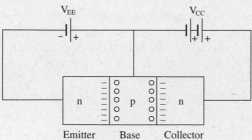

Fig. 7.18: Emitter-base junction is forward biased and base-collector junction is reverse biased.

The emitter current I_E is linked to the collector current I_C. Any increase in I_E would also increase I_C as I_B is small (Figure 7.19). I_E, I_B and I_C are conventional currents.

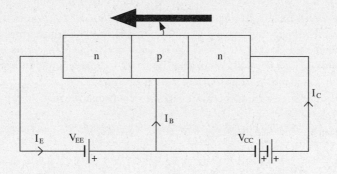

Fig. 7.19: A n-p-n transistor with bias voltages V_{EE} and V_{CC}.

The emitter current I_E is equal to the sum of the base current I_B and collector current I_C

$$I_E \quad = \quad I_B \quad + \quad I_C \qquad\qquad 7.2$$

By increasing the emitter-base voltage V_{EE} very slightly, a large change in I_E and I_C is made.

Usually the emitter-base voltage V_{EE} (0.7 V) is small but the base-collector voltage is much larger (6V) as shown in Figure 7.20. Most common transistors are n-p-n and are made of silicon. They are called **junction transistors** and are sometimes also called **bipolar transistors.**

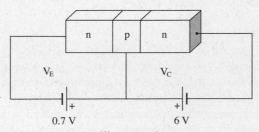

Fig. 7.20: Forward and Reverse bias voltages in silicon transistors.

A small current change between emitter and base can cause a 10–100 fold increase in base-collector current. As the emitter voltage is increased by a small amount, the emitter current rises and more electrons flow into the base. The base current I_B as well as the collector current I_C increases. The collector current is directly proportional to the base current in transistors:

$$I_C \quad \alpha \quad I_B$$
$$I_C \quad = \quad \beta I_B$$

The constant β **is the transistor current gain and equal to**

$$\beta \quad = \quad \frac{I_C}{I_B} \qquad\qquad 7.3$$

The transistor current gain β is the ratio of the collector current and the base current.

7.6 Transistor as an Amplifier

Amplification is important for practically all circuits. The transistor serves mainly as an amplifier.

Suppose an alternating voltage has to be amplified. This is connected to the emitter-base circuit. The output is taken across the load resistance R_L placed in the base-collector circuit. The input voltage is amplified (Figure 7.21).

As the emitter-base voltage changes with the AC voltage so would the emitter-base current change. The collector-base current is amplified β times. Thus, the collector-base voltage would also be amplified.

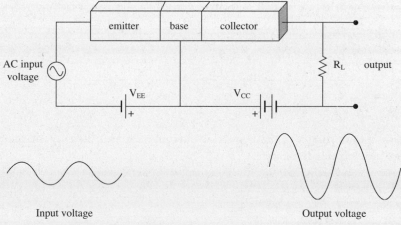

Fig. 7.21: A simple amplifier circuit. Also showing input and output AC waveforms.

To determine the voltage gain, consider Figure 7.22. The AC input voltage to be amplified is across the resistance R in the emitter-base circuit. Due to this AC voltage, the change in base current is equal to I_{in}. As the base current is amplified β times in the collector-base circuit because of the high reverse biasing V_{CC}, I_{in} is also amplified β times. The amplified I_{out} is equal to βI_{in}. The output voltage is across the load resistance R_L. R_L is a high resistance while the emitter-base circuit resistance is a comparatively small resistance R_{eb}.

The voltage gain of the amplifier is defined as

$$A_V = \frac{\text{output signal voltage}}{\text{input signal voltage}}$$

$$= \frac{V_{out}}{V_{in}}$$

But
$$V_{in} = I_{in} R_{eb}$$
$$V_{out} = I_{out} R_L$$

And
$$I_{out} = \beta I_{in}$$

Therefore
$$A_V = \frac{I_{out} R_L}{I_{in} R_{eb}}$$

$$= \frac{\beta I_{in} R_L}{I_{in} R_{eb}}$$

$$= \frac{\beta R_L}{R_{eb}} \qquad 7.4$$

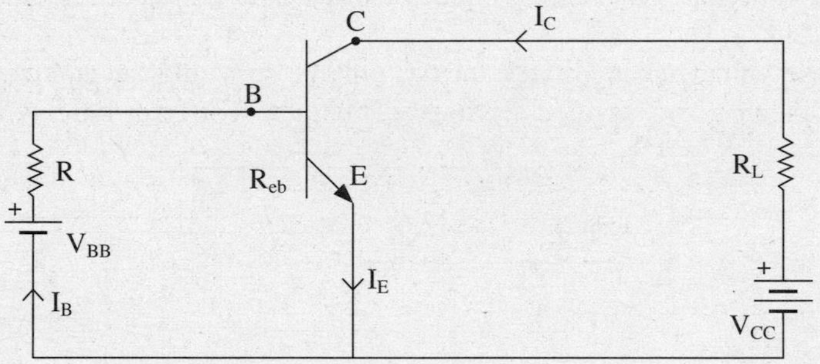

Fig. 7.22: A junction transistor amplifier circuit.

The voltage gain is the product of the transistor current gain and the ratio of the load resistance and the emitter-base circuit resistance.

To increase the voltage gain, the ratio of R_L/R_{eb} must be increased.

7.7 Transistor as a Switch

To grasp the working of the transistor as a switch, the change of the base input voltage with transistor output voltage needs to be understood.

As the base input voltage V_{in} increases, the output voltage remains nearly equal to the power supply voltage. As V_{in} increases beyond 0.7 V for silicon transistors, the output voltage falls dramatically (Figure 7.23).

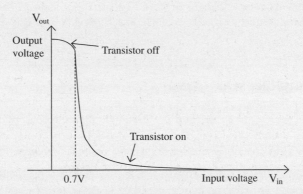

Fig. 7.23: Graph between base input voltage and transistor output voltage.

This abrupt fall in voltage makes the transistor suitable for working as a switch.

Consider the circuits of Figure 7.24.

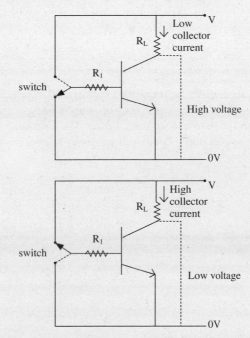

Fig. 7.24: The on and off states of a transistor switch circuit.

When the base current is small, the collector current is also low. The voltage drop across the load resistance R_L is low and across the transistor it is high. As the base voltage increases to above 0.7 V, the base current rises. The collector current is amplified many times. The voltage across load resistance

R_L increases and the voltage across the transistor decreases to a small voltage. **The transistor resistance seems to change from a high value to a low value.**

The transistor can be operated with the switch as shown but it is often used with a sensor. For instance, a temperature sensor or a thermistor could be used. As soon as the temperature reaches a certain level, the transistor switches on and an alarm is activated. When it is cold, the thermistor has a high resistance.

7.8 Operational Amplifier (Op - Amp)

Integrated circuits (ICs) which are an integral part of modern electronic circuitry are microelectronic circuits. On a single IC, a number of p-n junctions, transistors, capacitors, etc. may be connected. A Pentium microprocessor chip contains 7 million transistors and other electronic components. Individual electronic components are so small that a speck of dust can destroy it.

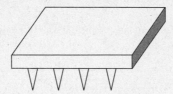

Fig. 7.25a: An 8-pin IC 741 chip.

The operational amplifier (op-amp) is an integrated circuit (IC). Operational amplifiers have a high gain typically of the order of 10^4 to 10^7. A standard op-amp is the 741. They are small chips with 8 pins, which are inexpensive and easy to use (Figure. 7.25a). The chip has more than a dozen transistors, half a dozen p-n junctions, capacitors and a number of resistances. Its schematic symbol and pin diagram are shown in Figure 7.25b.

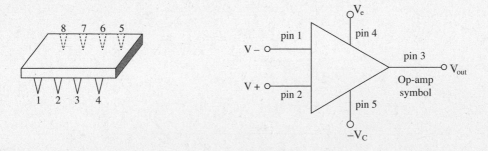

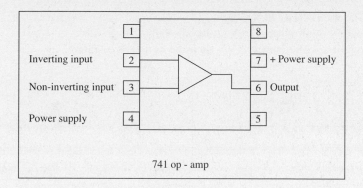

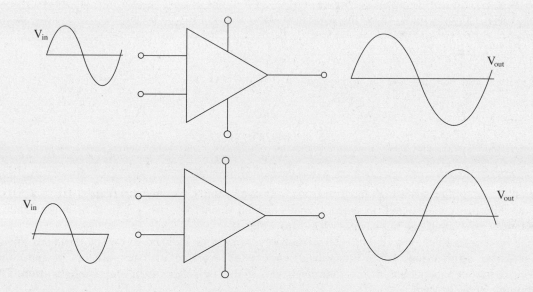

Fig. 7.25b: Schematic symbol and pin diagram of op-amp.

Pins 2 and 3 connect to op-amp inputs V_- and V_+, pin 6 connects to V_{out} and pins 4 and 7 connect to the power supply $\pm V_C$.

The op-amp is a differential amplifier. It amplifies the voltage difference of the two input signals V_+ and V_-. The input V_- is called the inverting input while V_+ is called the non-inverting input. The differential input voltage V_{id} is the difference of V_- and V_+

$$V_{id} \quad = \quad V_- \quad - \quad V_+ \qquad\qquad 7.5$$

The op-amp amplifier differential voltage V_{id} is amplified and not the input voltages separately. The output signal may be in phase with the input signal or 180° out of phase (Figure 7.26).

Fig. 7.26: Output signal in phase and out of phase with input signal.

The rule to determine the polarity of V_{out} is:

1. If V_+ is made positive with respect to V_-, then the output is positive.
2. If V_+ is made negative with respect to V_-, then the output is negative.

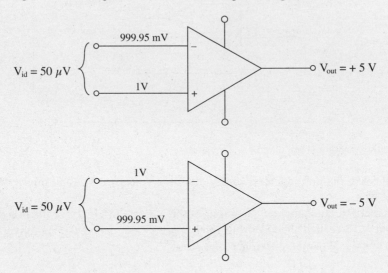

Fig. 7.27: Op-amp input and output polarity relation.

7.8.1 Op-amp Characteristics

The op-amp has a very high voltage gain of the order $10^4 - 10^7$. The maximum voltage gain is for an open-loop op-amp. It is defined as the ratio of the output voltage to the differential input voltage.

$$\text{Open-loop Voltage Gain} \quad = \quad A_{OL}$$

$$= \quad \frac{V_{out}}{V_{in}} \qquad\qquad 7.6$$

Supposing V_{id} is $\pm 50\ \mu V$ and the voltage gain A_{OL} is 100,000, then

$$V_{out} \quad = \quad (A_{OL})\,(V_{id})$$

$$= \quad (100,000)\,(\pm 50\ \mu V)$$

$$= \quad \pm 5\ V$$

Operational amplifier has a high input resistance. **It can amplify frequencies from 0 Hz to 1 MHz.** The amplifiers were designed to perform mathematical operations like subtraction, multiplication, integration and summation but have other uses as well.

The op-amp can be connected to external resistors, capacitors, etc. and so **can be designed to suit a particular application.** When operational amplifier is operated with no resistors or capacitors connected between its output and input terminals, it is said to operate in **open-loop configuration. The feedback circuit is open.**

Under this condition the op-amp has

a) A very high resistance R_{in} between its two input terminals.

If R_{in} is very high, no current is drawn between the input terminals.

b) The output resistance R_{out} is very low. $R_{out} \sim$ zero. The output voltage is independent of the load resistance.

In the op-amp, an input voltage of 1V cannot be amplified to 10^6 V but an input of 1 μV can be amplified to 1V. The **output voltage is limited by the bias or supply voltage** which generally is about ± 15 V.

Op-amp circuits use negative feedback mostly. In control systems feedback consists comparing the output of the system with the desired output and making a correction. Negative feedback involves the application of part of the output voltage to the inverting input. It is the process of coupling the output back in such a way as to cancel the input. This lowers the gain as part of the output is sent back to the input, but helps in reducing distortion and nonlinearity. The voltage gain is stable over a wide range of frequencies.

Op-amps are very reliable, small in size and give consistent results.

The op-amp has a number of applications, like linear constant gain amplifier, adder, subtractor, integrator, comparator.

The following applications will be discussed:

1) Inverting Amplifier
2) Non-inverting Amplifier
3) Comparator
4) Comparator as a Night Switch.

7.9 Inverting Amplifier

The inverting amplifier is a basic op-amp circuit. As it can be seen in Figure 7.28, it has an input resistance R_I and a feedback resistance R_F. V_o is the output voltage. The non-inverted terminal is grounded. It is at zero potential. V_{id} is the voltage between V– and V+ (ground). Since voltage gain A_{CL} for an op-amp is very high $(10^4 - 10^7)$, V_{id} is zero and V– is also at almost zero potential. It is said to be **at virtual ground**. The input signal V_{in} is so applied through input resistance R_I:

$$V_{in} \quad = \quad I\,R_I \qquad\qquad 7.7$$

As no current flows from the inverting to the non-inverting terminal, the current I through R_I then passes through R_F to the output terminal. The voltage across R_F is equal to the virtual voltage (0 V) minus V_o:

$$0 - V_o \quad = \quad -V_o \quad = \quad I\,R_F \qquad\qquad 7.8$$

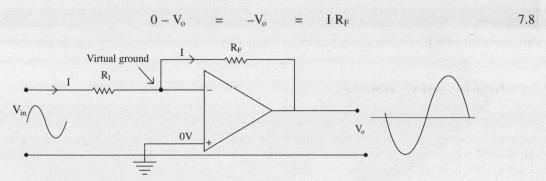

Fig. 7.28: An Inverting Amplifier.

Resistors R_I and R_F provide the negative feedback.

The voltage gain A_{CL} is the ratio of V_o to V_{in}

$$A_{CL} \quad = \quad \frac{V_{out}}{V_{in}} \qquad\qquad 7.9$$

where CL stands for closed loop.

Inserting equations 7.7 and 7.8 in equation 7.9 gives:

$$A_{CL} \quad = \quad \frac{-IR_F}{IR_I}$$

$$= \quad \frac{-R_F}{R_I} \qquad\qquad 7.10$$

The negative sign for closed loop gain shows that the output voltage polarity is opposite to the input voltage polarity. The gain depends on the ratio of the external resistances R_F and R_I only.

The op-amp has a very high gain. The gain is reduced by connecting a feedback resistor R_F. V_{out} of the inverted output voltage is fed back to the input. It cancels out the input voltage and so the input voltage is reduced.

Example: Find the voltage gain and output voltage in an inverting amplifier if

$$R_F \quad = \quad 10 \text{ k}\Omega, R_I \quad = \quad 1 \text{ k}\Omega \text{ and } V_{in} \quad = \quad 1 \text{ V}$$

Applying equation 7.10

The gain $\qquad\qquad\qquad A_{CL} \quad = \quad \dfrac{-10k\Omega}{1k\Omega}$

$$= \quad -10$$

And so output voltage $\qquad V_o \quad = \quad A_{CL} V_{in}$

$$= \quad (10) \ (1V)$$

$$= \quad -10 \text{ V}$$

The inverting amplifier as the name suggests has input and output signal 180° out of phase.

7.10 Non-Inverting Amplifier

In the non-inverting amplifier, the input and output signal is in phase. It has all the qualities of the inverting op-amp, except that it has higher input impedance. As the input voltages are equal, the differential voltage is approximately zero, so the input voltage V_{in} is across R_I and input current I (Figure 7.29) is given by

$$I \quad = \quad \frac{V_{in}}{R_I}$$

$$V_{in} \quad = \quad I\,R_I$$

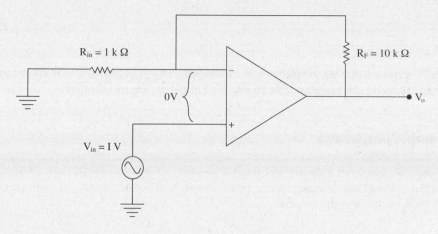

Fig. 7.29: A non-inverting closed loop op-amp circuit.

The current I flows through feedback resistance R_F to the output. The voltage across R_F is

$$V_F \quad = \quad I\,R_F$$

The output voltage V_o is taken with respect to ground. Therefore, V_o is the sum of V_I and V_F.

$$V_o \quad = \quad I\,R_F + I\,R_I$$

$$\quad = \quad I\,(R_F + R_I)$$

The gain A_{CL} is the ratio of V_{out} and V_{in}. A_{CL} is given by

$$A_{CL} \quad = \quad \frac{V_{out}}{V_{in}}$$

$$\quad = \quad \frac{I\,(R_F + R_I)}{I\,R_I}$$

$$\quad = \quad \frac{R_F}{R_I} + 1 \qquad\qquad 7.11$$

The gain of the non-inverting op-amp is greater than that of the inverting amplifier by 1.

Example: In a non-inverting op-amp $R_F = 10$ kΩ, $R_I = 1$ kΩ, $V_{in} = 1$ V. Find the output voltage.

$$V_{out} \;\; = \;\; \frac{10k\Omega}{1k\Omega} + 1$$

$$= \;\; 10 + 1$$

$$= \;\; 11V$$

Test Yourself: A non-inverting amplifier has an output voltage equal to 1.5 V. Its input resistance is 1.5 kΩ and its feedback resistance is 10.5 kΩ. What is its input voltage?

7.11 Op-amp Comparator

The two inputs of the open loop (no feedback) op-amp are put to use for using the op-amp as a comparator. An op-amp can compare the signal on one input with a reference voltage on the other. Figure 7.30 shows a comparator circuit.

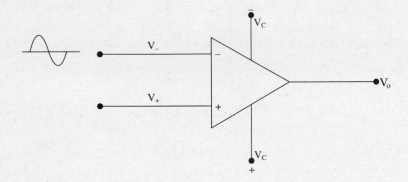

Fig. 7.30: Op-amp Comparator.

Power supply voltages in op-amp range from 5 to 15 V. The output voltage V_o is across the output terminal and ground.

$$V_o \quad \alpha \quad (V_- - V_+)$$

$$V_o \quad \alpha \quad A_{OL} (V_- - V_+)$$

$$A_{OL} \quad \alpha \quad \frac{V_o}{V_- - V_+}$$

The output voltage V_o is given by

$$V_o \quad = \quad A_{OL} (V_- - V_+)$$

Because of the high gain of the op-amp, a small input forces the output voltage to the saturation voltage. This saturation voltage cannot exceed the supply voltage $\pm V_{CC}$. The output voltage V_o will be $\pm V_{CC}$. The polarity depends on whether V_- is greater or less than V_+.

If $V_+ > V_-$ then V_o will be in phase to the input and equal to $+ V_{CC}$

And if $V_- > V_+$ then V_o will be 180° out of phase with the input and equal to $-V_{CC}$

Thus, by determining whether the output voltage is inverted or in phase, the two signals can be compared.

The op-amp non-inverting V+ can also be used as a reference voltage V_R and the signal to be compared is connected to the inverting input V−.

When $V- < V_R$, the output voltage $V_o = +V_{CC}$

And when $V- > V_R$, the output voltage $V_o = -V_{CC}$

7.12 Comparator as Night Switch

A circuit can be designed using the op-amp comparator to work as a light sensor. When the light intensity falls below a certain fixed value, the lights in streets or public places would be automatically switched on.

To the comparator are connected resistances R_1, R_2, R_3 and light dependent resistor R_L as shown in Figure 7.31. The resistance R_L depends on the light intensity. It is inversely proportional to the light intensity. During the day when there is more light R_L will be less. As the light intensity becomes less as night sets in, the resistance R_L will increase.

The voltage across R_2 is used as a reference voltage V_R and connected to the non-inverting input. It can be seen that V_R is equal to

$$V_R = \frac{R_2}{R_1 + R_2} V_{CC}$$

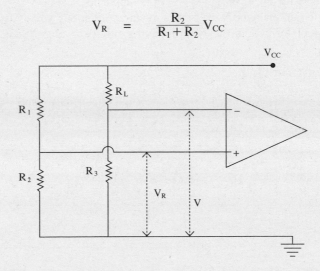

Fig. 7.31: The op-amp comparator switch circuit.

The Voltage V is the voltage which is to be compared to V_R. It is connected to V−. It will depend on the resistance R_L:

$$V = \frac{R_3}{R_L + R_3} V_{CC}$$

Voltage V alters with R_L. As R_L decreases, V will increase and vice versa.

As was the case in the comparator when

$$V > V_R \quad \rightarrow \quad V_o = -V_{CC}$$

and

$$V < V_R \quad \rightarrow \quad V_o = +V_{CC}$$

In the evening when it gets dark, V decreases, V_o becomes $+V_{CC}$. This triggers the switching system and the lights switch on. In the morning, V increases as R_L increases, V_o becomes $-V_{CC}$ and the lights switch off.

7.13 Digital Systems

Electronics involve circuits that have resistors, capacitors, p-n junctions, transistors, inductors and integrated circuits (ICs).

Some of these circuits have been incorporated into chips. It is not required to know how the circuits work. What is important is to know what a particular circuit does. Which input will produce which output?

One must know the difference between digital and analogue systems. For instance, the galvanometer and ammeter used in the physics laboratory use analogue system. All values of the voltage from 0–5 V, for example are shown on the meter. Digital system for this on the other hand, would operate at only two voltage levels: 0 V and 5 V. This is important because with the two values, the binary system using two digits 0 for 0 V and I for high voltage (5V) can be applied.

The AC waveform is an analogue waveform showing continuous values from −1 to 1 V while a square wave is an example of a digital waveform having only two values 0 and 2 V (Figure 7.32).

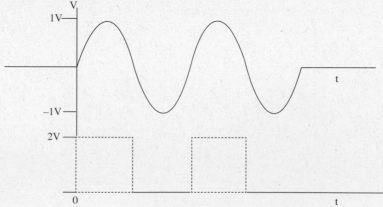

Fig. 7:32:   Top: Analogue waveform + 1 V to −1 V.
                Bottom: Digital waveform 0 V to 2 V.

The values low and high are 0 and 1 on the binary number system respectively. High voltages will be registered as 1 and very low voltages will be taken as 0.

Different logic circuits have been designed for specific purposes. These are known as **logic gates.**

7.14 Fundamental Logic Gates

The logic gates have numerous applications in everyday electronics. As it has been already mentioned, they are used in chips, in digital circuits known as flip-flops which have two stable states. They are also used in counters.

Logic gates are different circuits which give output of 1 or 0 for different inputs also as 1 or 0.

The three basic types of gates are:

1. AND gate
2. OR gate
3. NOT gate

7.14.1 AND Gate

The AND Gate can be compared to a circuit with 2 switches in series (Figure 7.33).

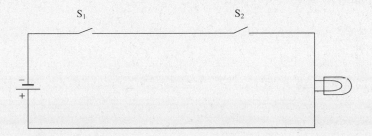

Fig. 7.33: A circuit equivalent to AND gate.

The bulb will only light up if **both switches are closed**.

The symbol for the AND gate is

A and B are the two inputs and Y is the output.

If both inputs are 1 then there will be an output of 1.

The **truth table** which describes all combinations of inputs and outputs of the AND gate is given in Table 7.1.

Table 7.1: Truth Table for AND Gate

Input		Output
A	B	Y
1	1	1
1	0	0
0	1	0
0	0	0

The Boolean algebra equation for the AND gate is

$$A.B \quad = \quad Y \qquad\qquad 7.12$$

A and B are inputs and Y is the output from the gate. The equation is read as

'A and B is equal to output Y'.

Example: For the given inputs 010010010 and 001010110 to a 2-input AND gate, find the output. Writing the inputs in waveforms gives

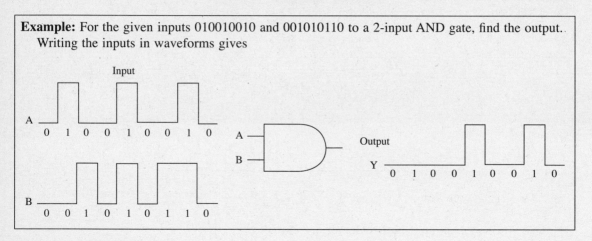

7.14.2 OR Gate

In a two input OR gate, if one or both inputs are 1, then the output is 1. A parallel switch electrical circuit can be visualized as an OR gate (Figure 7.34).

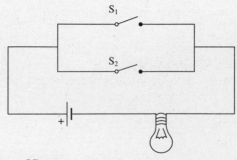

Fig. 7.34: A circuit equivalent to an OR gate.

If any one of the switches S_1 or S_2 is on, the bulb lights up.
The symbol for an OR gate is

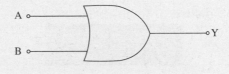

Its Boolean equation is   **A + B = Y**   7.13

It means **A or B will give an output Y.**
The truth table for a two input OR gate is shown in Table 7.2

Table 7.2: Truth Table for OR Gate

Input		Output
A	B	Y
1	1	1
1	0	1
0	1	1
0	0	0

Thus, the output is zero when there is no input in an OR gate.

Example: What is the output through a two input OR gate for inputs 0101 and 1101?
   Adding the outputs gives

$$\begin{array}{r} 0101 \\ +\ 1101 \\ \hline 1101 \end{array}$$

Both AND and OR gates can have more than 2 inputs.

7.14.3 NOT Gate

The NOT gate inverts the input. It has only one input and one output. If the input is 1, the output would be 0 and vice versa. In symbolic form, the output for the inverter whose input is A is written as $\bar{A}$. The symbol for a NOT gate is

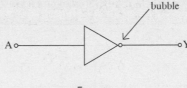

And the Boolean equation is $\bar{A} = Y$ 7.14

The bubble at the output of the gate means inversion. The NOT gate is also called inverter. The truth table is given in Table 7.3.

Table 3.3: Truth Table of NOT gate

Input A	Output Y
1	0
0	1

Example: If the input is 110010101, find the output for the NOT gate.

The input is	110010101
The output will be inverted	001101010

7.15 Other Logic Gates

Other gates have been designed that have specific logic.

7.15.1 NAND Gate – The Not AND Gate

As the name implies, the NAND gate gives the opposite output of that given by the AND gate. The symbol for the gate is

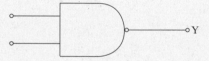

It consists of AND gate with a NOT bubble. This means that the output of AND gate should be inverted for the implementation of NAND gate. If both inputs are 1, the output will be 0 and if both inputs are 0, the output will be 1.

The Boolean equation is $\overline{A.B} = Y$ 7.15

The truth table is given in Table 7.4

Table 7.4: Truth table for 2-input NAND Gate

Input		AND Output	NAND Y
A	B		
1	1	1	0
1	0	0	1
0	1	0	1
0	0	0	1

The equivalent circuit for the NAND gate is shown in Figure 7.35. It comprises of a AND gate and a NOT gate. The output from AND gate is inverted.

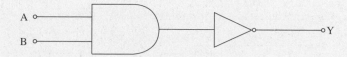

Fig. 7.35: An equivalent circuit for the NAND Gate.

7.15.2 NOR Gate (Not OR)
As the name suggests, the output of the OR gate is inverted to give the NOR output Y.

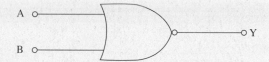

The Boolean equation is

$$\overline{A + B} = Y \qquad\qquad 7.16$$

Table 7.5 gives its truth table for 2 inputs

Table 7.5: Truth table for 2-input NOR Gate

Input		OR Output	NOR Y
A	B		
1	1	1	0
1	0	1	0
0	1	1	0
0	0	0	1

The NOR gate equivalent logic circuit is an OR gate in combination with a NOT gate as shown in Figure 7.36.

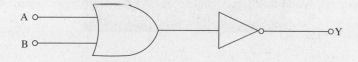

Fig. 7.36: An equivalent circuit for the NOR gate.

7.15.3 Exclusive OR (XOR) Gate

The exclusive OR (XOR) gate gives an output only when the number of inputs is odd. For an even number of inputs and when all inputs are 0, the output is 0. The gate symbol is

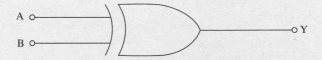

The truth table for a two input gate is given in Table 7.6.

Table 7.6: Truth table for 2-input XOR Gate

Input		XOR
A	B	Y
1	1	0
1	0	1
0	1	1
0	0	0

The Boolean equation is

$$A \oplus B \quad = \quad Y \qquad\qquad 7.17$$

or
$$\overline{A}\,B + A\,\overline{B} \quad = \quad Y$$

7.15.4 Exclusive NOR (XNOR) Gate

In the exclusive NOR, the output is high or 1 for even number of inputs equal to 1 and also when all inputs are zero. It gives an output of 1, when number of 1's at the input are even. It is an inverted XOR gate. The gate symbol is

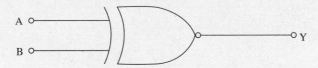

The truth table for a two input XNOR Gate is given in Table 7.7.

Table 7.7: Truth table for 2-input XNOR Gate

Input		XNOR
A	B	Y
1	1	1
1	0	0
0	1	0
0	0	1

The Boolean equation is

$$\overline{A \oplus B} \quad = \quad Y \qquad\qquad 7.18$$

or

$$\overline{\overline{A} B + A \overline{B}} \quad = \quad Y$$

Example: What is the output expression for the given circuit?

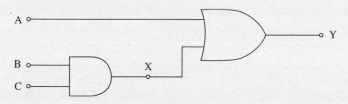

B and C are ANDed, so the output X is given by

$$B.C \quad = \quad X$$

X and A are ORed. The expression for Y is

$$X + A \quad = \quad Y$$
$$(B.C) + A \quad = \quad Y$$

Example: Draw the logic gate combination circuit for

$$A. \overline{B} + \overline{A}. B \quad = \quad Y$$

The first quantity A. B is a AND gate with inputs A and inverted input of B

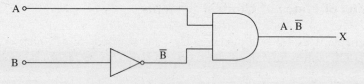

ORed to this is $\overline{A}$. B which is

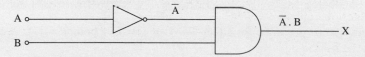

Combining X and X´ and ORing them gives Y

Example: (contd.)

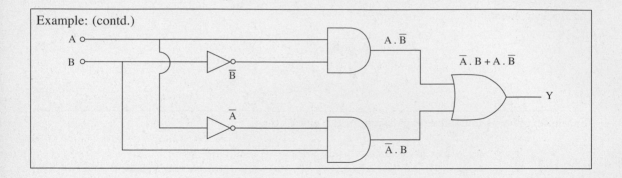

Example: If the inputs at A and B are 0 and 1 respectively, write the output state of NAND gate

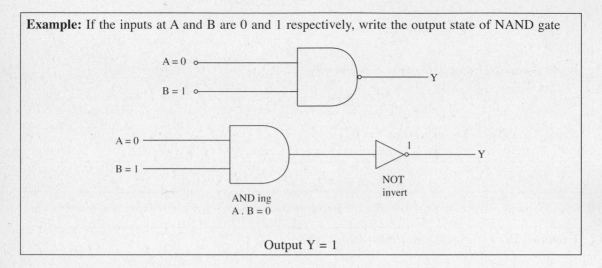

Output Y = 1

Test yourself: Draw the logic gates combination for (A. B) + (A + B) = Y.

7.16 Applications of Gates in Control Systems

Logic gates are particularly useful in designing control and sensor systems which would trigger alarms if certain parameters like pressure, temperature, concentration are exceeded beyond certain specified limits.

If a nuclear reactor system has to be constantly checked for pressure and concentration, then a circuit will have to be designed which would give a high or 1, when temperature (T) goes above the critical temperature and also when concentration (C) of the fissionable material exceeds a certain value. Below these particular values, the output is 0.

The alarm system will be activated for either T = 1 or C = 1 or both T = 1, C = 1. A OR gate with inputs T and C can be used:

$$T \quad + \quad C \quad = \quad A \text{ (alarm)}$$

The truth table is given in Table 7.8.

Table 7.8: Truth table for sensor system.

Input		Output
T	C	A
1	1	1
1	0	1
0	1	1
0	0	1

The schematic diagram for the system is shown in Figure 7.37.

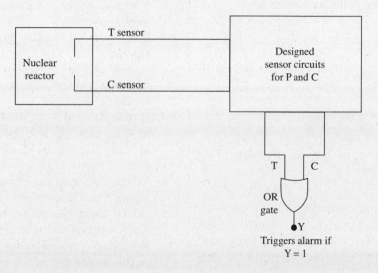

Fig. 7.37: A sensor system for a nuclear reactor using OR gate.

Summary

A semiconductor p-n junction allows current to flow in one direction when it is forward biased. The forward resistance is given by

$$R_F = \frac{\Delta V_F}{\Delta I_F}$$

The p-n junction is used for rectification. One semiconductor can be used for half-wave rectification while four p-n semiconductors are connected in a bridge arrangement for full-wave rectification.

The transistor is formed by two p-n junction placed back to back. Transistors are n-p-n and p-n-p. The emitter and base are forward biased while the base and collector are reverse biased. The emitter current I_E is equal to the sum of the base current I_B and collector current I_C

$$I_E = I_B + I_C$$

Transistor current gain β is a constant and equal to

$$\beta = \frac{I_C}{I_B}$$

Transistors are used mainly as amplifiers. The voltage gain is given by

$$A_V = \frac{\beta R_L}{R_{eb}}$$

Operational amplifiers (op-amps) are integrated circuits which are used most commonly. The differential input voltage V_{id} is amplified.

The open loop op-amp has no negative feedback. It has a very high gain A_{OL} and high resistance between its inverting and non-inverting inputs. In a closed loop inverting amplifier, the gain A_{CL} is given by

$$A_{CL} = \frac{-R_F}{R_I}$$

For a closed loop non-inverting amplifier gain is given by

$$A_{CL} = \frac{R_F}{R_I} + 1$$

The basic logic gates are AND, OR and NOT. Other gates are NAND, NOR, XOR, XNOR. These gate circuits are used in digital electronic circuits. Truth tables are used to study the input and output for the logic gates.

Questions

1. What charge carriers are present in the p-type zone, n-type zone and depletion zone?
2. What are the potential barriers of silicon and germanium semiconductor?
3. What is the definition of forward resistance?
4. Which amplifier has greater gain: inverting or non-inverting?
5. Can op-amps amplify both AC and DC signals?
6. List 3 characteristics of an operational amplifier? What is meant by a negative feedback?
7. If in an op-amp being used as a comparator $V_- < V_+$ then
   a) $V_o > -V_{CC}$ b) $V_o < -V_{CC}$
   c) $V_o = +V_{CC}$ d) $V_o = -V_{CC}$
8. Which logic gate has low output only when inputs A and B are high?
9. Which logic gate has a high output only when inputs A and B are low?
10. What does a bubble on the output of a logic circuit indicate?

Problems

1. If the collector current in a transistor is 15 mA and current gain is 100, what is the base current and the emitter current?
2. Find the collector current if base current is 0.5 mA and current gain is 200.

3. The open loop gain of op-amp is of the order of
   a) 10^2 b) 10^3 c) 10^4 d)10^5
4. In a non-inverting op-amp, the input voltage is 0.15 V, $R_F = 10$ kΩ and $R_I = 1.5$ kΩ, find A_{CL} and output voltage V_o.
5. Solve the above problem for an inverting amplifier.
6. In an inverting amplifier $R_F = 10$ kΩ, $R_I = 1$ kΩ and $V_{in} = 0.6$ V dc, calculate A_{CL} and V_o.
7. In the diagram given below, state which of the circuits is an op-amp comparator and which is an inverting amplifier?

   a)

   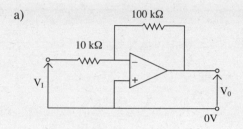

   b)

   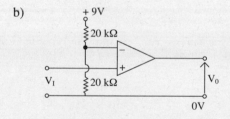

8. In circuit a) of problem 7, find the output voltage V_o, if V_I is 0.35 V.
9. Draw the logic symbol and construct the truth table for a 2-input NAND gate.
10. Draw the logic symbol and construct the truth table for a 2-input NOR gate.
11. Two electrical signals represented by A=101101 and B=110101 are the inputs into a 2-input AND gate. Sketch the output.
12. An electrical signal is expressed as 101011. If this signal is applied to a NOT gate, what would be the output signal?
13. The output of NOR gate is 1 when
    a) all inputs are zero
    b) at least one input is zero
    c) all inputs are one
    d) all of these
14. Identify the logic gate symbol and write the Boolean equation. If A = 1 and B = 0, write down the output state.

    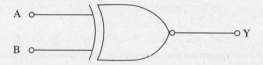

15. What is the output for a 2-input NOR gate, if A = 1 and B = 0?
16. What is the output for a 2-input XOR gate, if A = 1 and B = 1?

8 Modern Physics

OBJECTIVES

After studying this chapter, the student should be able:
- to understand the special theory of relativity.
- to understand the black body radiation and the concept of emissivity.
- to understand the energy-temperature distribution curve.
- to understand Planck's assumption and explanation for the energy distribution.
- to understand the particle nature of light and its consequences: Photoelectric Effect, Compton Effect and Pair Production.
- to understand Davisson and Germer experiment showing particle wavelengths.
- to understand the applications of photoelectric effect.
- to understand the Compton effect and scattering of photons.
- to understand the energy requirement for
   Pair production
   Pair annihilation
- to understand de Broglie's hypothesis showing that particles have wavelengths (wave quantities).
- to understand the electron microscope.
- to understand Heisenberg's Uncertainty Principle for measurement of position and momentum, energy and time accurately at the same instant of time.

8.1 Relative Motion

It has already been discussed in Physics for Class IX that a frame of reference is necessary for making measurements as motion is relative.

It is important to define this frame of reference because the motion of a car with respect to the surface of the earth is different to what it is with respect to the sun. If we consider the surface of the earth as the frame of reference, the motion is as we normally watch it drive away. But if we look at the motion from the sun as the reference frame, then the revolution as well as the spin of the earth and the velocity of the car would have to be taken into account to figure out the motion (Figure 8.1 a, b).

velocity

Fig. 8.1a: Reference: surface of the earth.

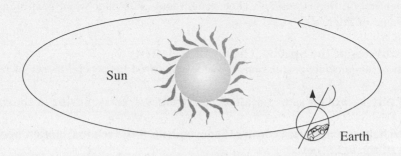

Fig. 8.1b: Reference: Sun.

For you, your television set is stationary but for a person looking at it from the center of the earth, it would appear to move in a circle because of the spin of the earth. A woman sitting in a moving train is stationary relative to the train but would be moving relative to the tree outside.

There is **no absolute motion.** For absolute motion, a point would have to be found which is stationary and unchanging. Such a point does not exist in our present day knowledge of the universe. Every known body in the universe is in motion. The universe is expanding. There is no universal frame of reference that can be applied everywhere in the universe.

Motion is relative and a frame of reference has to be selected relative to which measurements can be made. A frame of reference is a coordinate system from which measurements are made.

8.2 Inertial Frame of Reference

An inertial frame of reference is a frame of reference moving with uniform velocity.

It is not possible to detect constant velocity motion of an object. Suppose a man is sealed inside a car. The car is moving at constant velocity on a perfectly smooth and straight, horizontal road. There is no experiment or method that can be performed in the car that can determine whether it is moving or at rest. Imagine the car is at rest. The man drops a ball. It falls on his foot. When the car is moving with constant velocity and he repeats the experiment, the motion of the ball will appear exactly the same and it will hit his foot.

But if the car accelerates (increases speed or changes direction), the ball's motion will be different and it will not fall on his foot. Thus, accelerated motion can be detected.

The special theory of relativity which will be discussed in the next section is based on **inertial reference frames** while the general theory deals with accelerated frames of reference.

8.3 Special Theory of Relativity

Einstein's Special Theory of Relativity was able to explain certain phenomena which Classical and Newtonian theories were unable to explain.

The amazing fact is that Einstein published his special theory of relativity when he was just 26 years old in 1906. In fact, the theory of relativity completely altered our way of thinking and paved the way to modern physics. The special theory of relativity connects mass and energy, time and space formerly believed to be entirely different entities. Time, length and mass alter when measurements are made from another frame of reference.

8.3.1 The Postulates of the Special Theory of Relativity

Einstein gave two fundamental postulations on which he based his special theory of relativity. These are:

1. **The laws of physics are the same for all inertial frames (frames moving at constant velocity to each other).**
2. **The speed of light is a universal constant independent of the relative motion between observer and motion of the light source.**

Light has a speed of 3×10^8 m/s. It is denoted by c.

The first postulate implies that two observers in different frames of reference measuring an event in a third frame will find the same laws of physics like Newton's laws and law of conservation of energy valid. It is because of this postulate that a person moving with constant velocity is unable to detect motion.

On the basis of the postulates of the General Theory of Relativity:

1) **A time interval** in a moving frame of reference appears to be extended or dilated to a **stationary observer.**
2) The **length** measured in a moving frame of reference appears to be contracted as measured by a stationary observer.
3) The **mass** in a moving frame of reference appears more as measured by an observer from a stationary frame.

These values become significant only when the velocities approach the velocity of light ($c = 3 \times 10^8$ m/s). For ordinary speeds up to the order 10,000 m/s, the difference between Newtonian mechanics and values calculated on the basis of Einstein's theory can be ignored.

8.3.2 The Light Clock

Two mirrors M_1 and M_2 are placed in a spaceship as shown. The spaceship is at rest relative to observers A and B (Figure 8.2a).

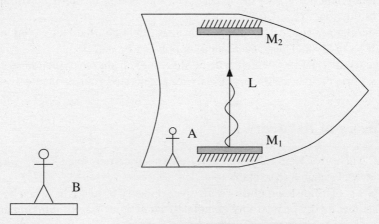

Fig. 8.2a: Spaceship stationary with respect to A and B.

Light is flashed from mirror M_1. It hits M_2 and is reflected back. As soon as it strikes M_2, let the time be equal to a unit of time t_o. If the distance between the mirrors is L then

$$t_o \quad = \quad L/c \qquad\qquad \text{as } v = S/t$$
$$\qquad\qquad\qquad\qquad\qquad c = \text{velocity of light}$$
$$L \quad = \quad c\, t_o$$

The spaceship now moves with a velocity **v**. Let the time it takes for the light to go from M_1 to M_2 be t.

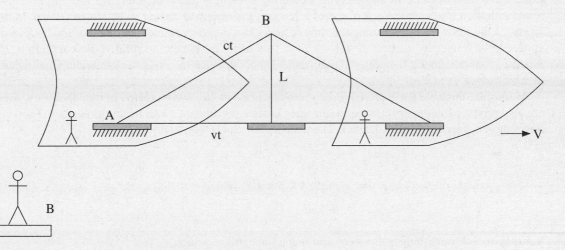

Fig. 8.2b: Spaceship moving with velocity **v** with respect to observer B.

For observer A sitting in the spaceship and moving with it, light will again appear to go vertically upwards as in Figure 8.2a. The observer B sees the light as shown in Figure 8.2b. As the spaceship moves to the right with velocity **v**, the light will go along AB. In time t the distance covered by the light is ct and by the spaceship the distance covered will be vt.

Using Pythagoras Theorem

$$(ct)^2 \quad = \quad L^2 + (vt)^2$$
$$(ct)^2 \quad = \quad (ct_o)^2 + (vt)^2$$
$$c^2 t_o^2 \quad = \quad t^2 (c^2 - v^2)$$
$$t_o^2 \quad = \quad t^2 (1 - v^2/c^2)$$
$$\mathbf{t_o} \quad = \quad t\sqrt{(1 - v^2/c^2)} \qquad\qquad\qquad 8.1$$

Equation 8.1 relates the time t_o when the spaceship (frame of reference) is at rest with respect to the observer B and the time t when it is in constant relative motion to the observer B who is measuring the time. This relationship between t_o and t is given in equation 8.1.

8.3.3 Time Dilation

It is seen that the time t measured by an observer in relative motion to the frame of reference in which he is making the measurement is

$$t = \frac{t_o}{\sqrt{(1 - v^2/c^2)}}$$

As
$$\sqrt{(1 - v^2/c^2)} < 1$$

$$\mathbf{t} > \mathbf{t_o} \qquad\qquad 8.2$$

Equation 8.2 shows that to an observer measuring time in a frame of reference moving at constant relative motion to the observer's frame of reference, the time will appear to be slowed down. In other words, if the observer is measuring the time of a clock in the other frame, the minutes and hours would appear to be longer by a factor of $\sqrt{(1 - v^2/c^2)}$. This factor is termed the **relativistic correction**. Here v is the relative velocity between the two frames—one in which the observer is standing and the other in which clock is placed.

Muons are unstable particles that decay into electrons in $2\mu s$ when created in the laboratory. Muons are also produced in the upper atmosphere when cosmic rays interact with the air molecules. They move down at very fast velocities (0.998c). The distance in which they should decay is

$$S = v\,t$$
$$= (0.988 \times 3 \times 10^8 \text{ m/s})(2 \times 10^{-6} \text{ s})$$
$$\sim 600\text{m}$$

Yet they travel much longer distances and reach the earth's surface.

$$\text{Applying} \quad t = \frac{t_o}{\sqrt{(1 - v^2/c^2)}}$$

$$= \frac{2\mu s}{\sqrt{1 - \dfrac{0.998c^2}{c^2}}}$$

$$= 31.6 \ \mu s$$

The fast velocity muons have much longer lifetimes than the muons produced in the laboratory which are at rest.

Please note measurements are being made by a stationary observer at rest on the surface of the earth.

Example: In an atom at rest in the laboratory an event takes place in 10^{-6} s. How much time will this event require if this atom is moving with a speed of 5×10^7 m/s as measured by an observer in the laboratory.

$$t_o = 10^{-6} \text{ s}$$
$$v = 5 \times 10^7 \text{ m/s}$$
$$c = 3 \times 10^8 \text{ m/s}$$

Example: (contd.)

Using

$$t = \frac{t_o}{\sqrt{1 - \dfrac{v^2}{c^2}}}$$

$$= \frac{10^{-6}\,s}{\sqrt{1 - \dfrac{(5 \times 10^7)^2}{(3 \times 10^8)^2}}}$$

$$= \frac{10^{-6}\,s}{\sqrt{1 - \dfrac{5^2}{(30)^2}}}$$

$$= \frac{10^{-6}\,s}{\sqrt{1 - \dfrac{(1)^2}{(6)^2}}}$$

$$= \frac{10^{-6}\,s}{\sqrt{1 - (0.166)^2}}$$

$$= \frac{10^{-6}\,s}{\sqrt{1 - 0.027}}$$

$$= \frac{10^{-6}\,s}{\sqrt{0.97}}$$

$$= \frac{10^{-6}\,s}{0.98}$$

$$= 1.02 \times 10^{-6}\,s$$

$$= 1.02\ \mu s$$

8.3.4 Length Contraction

Again if length is measured in two frames moving with constant relative velocity to each other (Figure 8.3), it will be observed that the meter rod would appear to be shorter (L) when measured by the observer B. For observer A who is in the same frame of reference as the meter rod, the length will be equal to L_o.

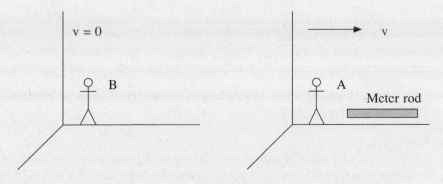

Fig. 8.3: Observer B measuring metre rod in frame of reference of A.

Observer B would measure the meter rod and it would appear shorter. The result would be

$$L = L_0 \sqrt{(1 - v^2/v^2)} \qquad\qquad 8.3$$

This contraction in length as measured from one frame moving with constant relative velocity to another frame is called **Lorentz contraction**.

Example: A spacecraft is moving at a speed v. Its length measured from earth is 99% of its length when it was at rest on earth. What is its speed?

As the measured length l is 99% of the length at rest l_0

Therefore

$$l_0 = 100 \text{ units}$$

$$l = 99 \text{ units}$$

Then

$$l = l_0 \sqrt{(1 - v^2/c^2)}$$

$$99 = 100 \sqrt{(1 - v^2/c^2)}$$

Squaring

$$\frac{99^2}{100^2} = 1 - v^2/c^2$$

$$\frac{v^2}{c^2} = 1 - (0.99)^2$$

$$= 1 - 0.98$$

$$= 0.02$$

$$v^2 = (0.02)\, c^2$$

$$v = \sqrt{(0.02)}\, c$$

$$= (0.14) \times (3 \times 10^8 \text{ m/s})$$

$$= 0.42 \times 10^8 \text{ m/s}$$

$$= 4.2 \times 10^7 \text{ m/s}.$$

Test yourself: By what percentage will a body's length be shortened from its length at rest if it is moving at 9/10 of the velocity of light?

8.3.5 Mass Variation

The mass of a body is not constant although in classical physics it was taken as a constant. The mass depends on the body's velocity with respect to the observer. The faster the mass moves, the more mass it has. We do not see a change in the mass with the masses we deal with everyday. But the variation of mass becomes significant when dealing with very small mass on the sub-atomic scale like electrons and protons which also move with speeds approaching that of light.

The mass m of a body moving with velocity **v** relative to the observer measuring it is given by

$$m = \frac{m_o}{\sqrt{(1 - v^2/c^2)}} \qquad\qquad 8.4$$

$$m > m_o \qquad\qquad \text{as } \sqrt{(1 - v^2/c^2)} \text{ is less than 1} \qquad 8.5$$

Here m_o is the mass when the body is at rest and is known as the **rest mass**.

For example, let us compare the mass change of a spacecraft moving in space to that of an electron:

A spacecraft's velocity is 11000m/s. If its mass is 60000 kg, the mass m at this speed can be determined using equation 8.4:

$$m = \frac{m_o}{\sqrt{(1 - v^2/c^2)}}$$

$$= \frac{60000 \text{ kg}}{\sqrt{[1 - (11000\text{m/s})^2/(3 \times 10^8 \text{m/s})^2]}}$$

$$\sim 60000.0000006 \text{ kg}$$

The spacecraft's mass is now 0.0000006 kg more, which is hardly any significant change in the spacecraft's mass and can be ignored.

On the other hand, the mass m for an electron whose rest mass m_o is 9.109×10^{-31} kg and velocity 0.998c can be computed

$$m = \frac{m_o}{\sqrt{(1 - v^2/c^2)}}$$

$$= \frac{(9.109 \times 10^{-31} \text{ kg})}{\sqrt{[1 - (0.998c)^2/c^2]}}$$

$$= 144.098 \times 10^{-31} \text{ kg}$$

The mass has increased nearly 16 times.

The above calculations show that the increase in mass for large bodies moving at speeds of the order of 10^4 m/s is insignificant. The laws of classical physics remain valid. Thus, mass in classical physics can be considered a constant. But for sub-atomic systems which are moving with speeds approaching the speed of light, the relativistic correction to mass must be applied.

8.3.6 Mass-Energy Relation

In nuclear reactions as well as in fusion reactions, huge amounts of energy are released and a detectable change of mass is observed.

Energy and mass that were thought to be conserved separately are interchangeable. The mass-energy relationship, an outcome of the theory of relativity is given by

$$E = mc^2 \qquad\qquad 8.6$$

where E is the total energy.

When $v = 0$ and the object is at rest, the total energy is termed **rest mass energy E_o**.

$$E_o \quad = \quad m_o c^2 \qquad (v = 0, \, m = m_o) \qquad\qquad 8.7$$

The rest mass energy E_o is the energy that is equivalent to the mass of the body at rest.

Even in chemical reactions, mass in converted into energy but it is so very minute that it goes undetected. When a body is moving with velocity **v**, it has a total energy E equal to the sum of its rest mass energy and kinetic energy, KE:

$$\mathbf{E} \quad = \quad \mathbf{m_o c^2 + KE} \qquad\qquad 8.8$$

$$\mathrm{KE} \quad = \quad E - m_o c^2$$

$$= \quad mc^2 - m_o c^2$$

$$= \quad \mathbf{\Delta m \ c^2} \qquad\qquad 8.9$$

The kinetic energy is equal to the increase in mass times (velocity)^2.

Another useful equation for the total energy is

$$\mathbf{E^2} \quad = \quad \mathbf{(m_o c^2)^2 + p^2 c^2} \qquad\qquad 8.10$$

where KE has been written in terms of the relativistic momentum p.

If the particle is at rest $(v = 0)$

$$E^2 \quad = \quad (m_o \, c^2)^2$$

$$E \quad = \quad m_o \, c^2$$

and if it is a particle that has no mass (photon)

$$\mathbf{E^2} \quad = \quad \mathbf{p^2 c^2} \qquad \text{or} \qquad \mathbf{E = p \ c} \qquad\qquad 8.11$$

8.3.7 NAVSTAR Navigation System

The NAVSTAR Navigation System uses the relativistic correction $\sqrt{(1 - v^2/c^2)}$ to determine the variation in the Doppler's shift. A number of such satellites are coursing around the earth. They are able to predict an aircraft's speed to an accuracy of 1.4 cm/s. Thus, where the aircraft would be in one hour can be predicted within a distance of 50 m. Without this relativistic correction, the aircraft's position can only be predicted within an accuracy of 760 m.

The principle of relativity follows directly from an absence of a universal frame of reference. If the laws of physics were different for different observers in relative motion, they could determine which of them were stationary in space and which were moving.

8.4 Black Body Radiation

All bodies with temperatures more than the absolute zero, 0 Kelvin (K), radiate electromagnetic radiations. When the temperature of the object is low, longer wavelengths are emitted. At higher temperatures, shorter wavelengths at higher frequencies are emitted. The white-hot filament of the bulb emits visible light at 1700 K and the red-hot heater wire is at 1000 K. We, too, at 310 K emit infrared radiation.

A black body absorbs all radiation that falls on it. It is a perfect absorber of all radiations falling on it. A black body absorbs not only infrared and visible, but all other radiations falling on it. Black velvet and lampblack are near ideal blackbodies. A box with a small hole approximates a black body (Figure 8.4). Any radiation that enters this box is continuously reflected from the inside walls until it is completely absorbed. There is a very small possibility that it will escape out.

Objects both absorb and radiate energy. The rate of absorption and emission may not be equal. At the same temperature, some bodies may have higher rate of emission radiation than others. The emissive rate is measured by the **emissivity.**

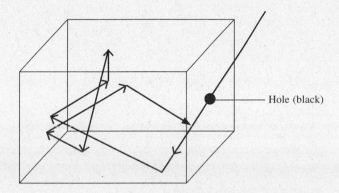

Fig. 8.4: A box with a small opening through which radiation is unlikely to escape is an ideal black body.

The emissivity is defined as the ratio of the emission power of the object at a temperature T to the emission of a black body at T.

$$\textbf{Emissivity} \quad = \quad \frac{\text{Emission power of object at T}}{\text{Emission power of black body at T}} \qquad 8.12$$

The emissivity has a range 0 to1. The emissivity depends on the surface of the body—its nature and the material it is made up of. For the same temperature, some bodies may radiate more than others. Soot has an emissivity of 0.95.

An ideal body that has a maximum emissivity of 1 is called a Black Body. A black body is a perfect absorber and a perfect emitter.

8.4.1 Stefan – Boltzmann Law

Stefan and Boltzmann independently studying radiations from objects found experimentally and theoretically that the amount of radiation emitted is proportional to the body's surface area and the absolute temperature respectively.

$$E \quad \alpha \quad AT^4$$
$$= \quad \sigma AT^4 \qquad\qquad 8.13$$

where $\sigma = 5.67 \times 10^{-8}$ W/(m²K⁴) is the constant of proportionality and is known as the **Stefan-Boltzmann constant.**

8.4.2 Radiation Spectrum—Intensity Distribution of Radiation

It is observed in the emission spectrum of black bodies at different temperatures that the wavelength at which maximum radiation occurs shifts towards the shorter wavelengths at higher temperatures (Figure 8.5).

$$\lambda_{max} \quad \alpha \quad 1/T$$
$$\lambda_{max} T \quad = \quad \text{constant} \qquad\qquad 8.14$$
$$= \quad 2.898 \times 10^{-3} \text{ m K}$$

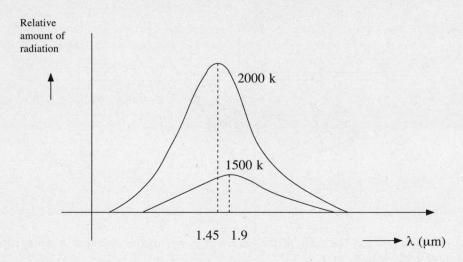

Fig. 8.5: Energy distribution curves for black bodies at 2000 K and 1500 K.

This is known as **Wein's Displacement Law.**

It can be observed from these energy distribution curves that the energy is not uniformly distributed. For a given temperature at certain wavelengths, radiation is maximum. The maximum wavelength shifts towards the shorter wavelength at higher temperatures.

The radiated energy is proportional to the area and (temperature)⁴.

Example: The maximum energy from the sun is radiated at 0.5 μm. What is the sun's temperature?

$$\lambda_{max} \quad = \quad 0.5 \ \mu m \quad = \quad 0.5 \times 10^{-6} \ m$$

$$T \quad = \quad ?$$

Applying Wein's Displacement Law (equation 8.14)

$$\lambda_{max} \ T \quad = \quad 2.898 \times 10^{-3} \ m.K$$

$$T \quad = \quad \frac{2.898 \times 10^{-3} \, m.K}{0.5 \times 10^{-6} \, m}$$

$$= \quad 5.796 \times 10^{3} \ K$$

$$= \quad 5796 \ K$$

8.4.3 Planck's Explanation

Classical physics is unable to explain the energy distribution spectra. Planck formulated his theory in 1900. He equated a black body to many atomic oscillators. He also observed that the radiation was not continuous, but in the form of minute bundles of energy called **quanta.** The oscillators could only emit discrete quantities of energy given by

$$E = n \, h \, \upsilon \qquad\qquad n = 1, 2, 3, \dots\dots \qquad\qquad 8.15$$

Each quantum has energy equal to

$$E = h \, \upsilon \qquad\qquad 8.16$$

where **h is the Planck's constant and is equal to 6.63×10^{-34} J.s.**

8.4.4 The Photon

Using the concept of the quanta, Planck was able to explain the black body spectra. A quanta of light energy is called a **photon.** Light is regarded as bundles of photons. A bulb would radiate many different frequencies but for each frequency the photon is given by

$$E \quad = \quad h \, \upsilon$$

$$= \quad hc/\lambda \qquad\qquad as \ c = \lambda \, \upsilon \qquad\qquad 8.17$$

Equating E = pc (equation 8.11), equation 8.17 becomes:

$$pc \quad = \quad hc/\lambda$$

$$p \quad = \quad h/\lambda \qquad\qquad 8.18$$

As it can be seen from the equation, $E = h\upsilon$ for radiations of smaller frequencies (υ) like radio waves the energy for each quantum will be extremely small. Thus, the spectrum would appear to be continuous.

For x-rays and γ rays, the quanta have much higher values of energy due to x-rays and γ rays' very high frequencies. They are detected easily.

Example: What would be the smallest amount of energy emitted by green light?
($\upsilon = 6 \times 10^{14}$ Hz, $\lambda = 5 \times 10^{-7}$m)

$$E = h\upsilon$$
$$= (6.63 \times 10^{-34} \text{ J s})(6 \times 10^{14} \text{ Hz})$$
$$\sim 4 \times 10^{-19}\text{J}$$

8.5 Interaction of Electromagnetic Radiation with Matter

In this section, we will see properties of light that show its particle behaviour as opposed to the wave theory of light. The photoelectric effect, compton's effect and pair production, all support the photon (particle) model. These phenomena will now be discussed.

8.5.1 Photoelectric Effect
In 1887, the photoelectric effect was discovered by Heinrich Hertz, a German physicist. When light falls on a metal plate, electrons are emitted. They are called **photoelectrons.**

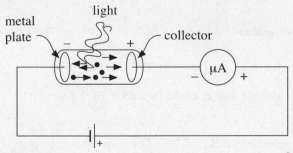

Fig. 8.6: Production of photoelectrons.

Air is evacuated from a glass bulb in which a metal plate and a collector are placed. The collector (anode) is made positive (battery) to attract the photoelectrons being emitted by the metal plate (cathode) which is illuminated by light. A galvanometer or micro-ammeter is placed in the circuit to measure the photoelectric current (Figure 8.6). The photoelectrons are so called because they are produced by light (photo) illuminating the metal plate. Light carries energy. Energy is absorbed by the metal. Certain electrons acquire enough kinetic energy to leave the metal. This explanation seems logical but when examined in detail, interesting facts emerge which cannot be explained by classical physics.

1. One such fact is that even if the light beam barely illuminates the metal photoelectrons emerge instantaneously in time $< 10^{-9}$ s after the metal surface is illuminated. One would imagine that a certain amount of energy needs to be accumulated by the electron before it can exit the metal and so the process should take some time.

2. The photoelectron current stops when the light is blocked proving that the electrons are produced by light.
3. When the intensity of illumination is increased, the number of photoelectrons increases (photoelectric current), but they do not move faster (Figure 8.7). They do not acquire more energy but their numbers increase.

$$\text{No. of electrons emitted} \quad \alpha \quad \text{light intensity} \qquad 8.19$$

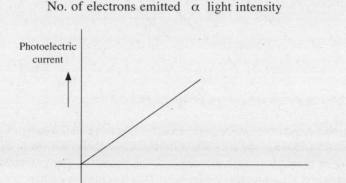

Fig. 8.7: As light intensity increases, the photoelectric current increases.

The classical electromagnetic theory predicts that as light intensity is increased, the energy of the photoelectrons should also increase.

4. The kinetic energy of the photoelectrons unexpectedly is found to be proportional to the frequency υ of the incident light (Figure 8.8).

$$\text{KE}_{\text{photoelectrons}} \quad \alpha \quad \upsilon_{\text{light}} \qquad 8.20$$

5. At frequencies below a certain critical value, photoelectrons are not produced, however intense the light source may be (Figure 8.8a). This frequency is characteristic of the metal used and is called the threshold frequency. For metal A, the threshold frequency is υ_A and for metal B it is υ_B. Below these frequencies, no photoelectrons are emitted. Figure 8.8b gives the threshold frequencies of cesium, potassium and sodium. The kinetic energy of the photoelectrons depends on the frequency of light.

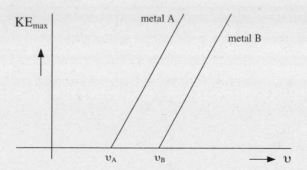

Fig. 8.8a: Graph between KE_{max} of the photoelectrons and frequency υ of incident light.

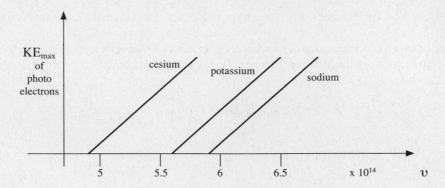

Fig. 8.8b: The threshold frequencies of cesium, potassium and sodium.

The energy of the photoelectrons can be computed by reversing the voltage of the battery. This will repel the photoelectrons moving towards the collector. Only those electrons will reach the collector whose kinetic energies are high enough to overcome the potential. The potential is now increased till no current is registered on the ammeter (Figure 8.9). This potential is known as the **stopping potential** V_{stop}.

The maximum kinetic energy of the fastest photoelectron is given by

$$\tfrac{1}{2} \, m v^2_{max} \quad = \quad e \, V_{stop} \qquad\qquad 8.21$$

8.5.1.1 Explanation of Photoelectric Effect
Using Planck's idea of quanta, Einstein was able to come up with a plausible explanation for the experimental data.

1. Einstein photoelectric equation is

$$\mathbf{h\upsilon} \quad = \quad \mathbf{\Phi_o} \quad + \quad \mathbf{KE_{max}} \qquad\qquad 8.22$$

$$\mathbf{h\upsilon} \quad = \quad \mathbf{h\upsilon_o} \quad + \quad \mathbf{KE_{max}} \qquad\qquad 8.22a$$

| Energy of photon | = | Work function | + | Maximum energy of photoelectron |

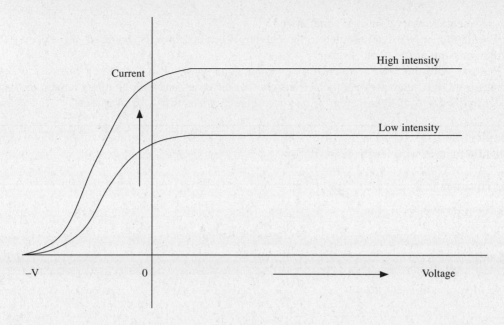

Fig. 8.9: Graph showing the stopping potential equal to –V.

The high photon energy $h\upsilon$ is acquired by the electron. **The minimum energy it requires to leave a certain metal is $h\upsilon_0$. It is called the work function Φ_0 of the metal.** The rest of the energy appears as the maximum KE of the photoelectron. The threshold frequency is the minimum frequency of the light that can emit photoelectrons.

2. This also explains why photoelectrons are emitted immediately. They acquire a quantum of energy $h\upsilon$. If this energy is equal or more than the work function $h\upsilon_0$ for that particular metal, they would leave the metal surface. The extra energy appears as the electron's kinetic energy (Figure 8.10).
3. Electrons may lose some of the energy in collisions and so they have different kinetic energies.
4. More intense light means more photons. The number of photoelectrons increases with the intensity of light (Figure 8.7).

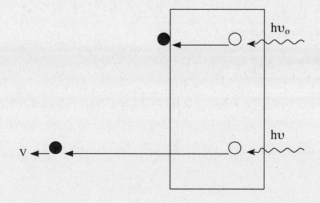

Fig. 8.10: Showing two photoelectrons, one produced by photon of $E = h\upsilon_0$ and the other by photon of energy $h\upsilon > h\upsilon_0$.

The experimental observations are summarized.

1. The light energy is concentrated in the photon. When a photon is absorbed, there is no delay in the emission of the photoelectron.
2. Photons with the same frequency υ have the same energy $E = h\upsilon$. When intensity of light is increased, the number of photoelectrons emitted increases but their energies remain unchanged.
3. As frequency υ of light increases, energy of the photoelectron also increases.

Example: The work function for silver is 4.2 eV. What is the kinetic energy of the photoelectrons emitted by light of wavelength 2.5×10^{-7} m?

Using equation 8.22	KE	$=$	$h\upsilon - h\upsilon_o$
Work function	$h\upsilon_o$	$=$	4.2 eV
		$=$	$4.2 \times 1.6 \times 10^{-19}$ J
		$=$	6.7×10^{-19} J

$$h\upsilon = h\,(c/\lambda) = \frac{(6.63 \times 10^{-34}\,\text{J.s})\,(3 \times 10^{8}\,\text{m/s})}{2.5 \times 10^{-7}\,\text{m}} \qquad c = \upsilon\lambda$$

		$=$	7.956×10^{-16} J
		$=$	7956×10^{-19} J
But	KE	$=$	$h\upsilon - h\upsilon_o$
		$=$	$(7956 - 6.7) \times 10^{-19}$
		$=$	7949.3×10^{-19} J

Example: Copper has a threshold frequency of 1.1×10^{15} Hz. Light of frequency 2.5×10^{15} Hz falls on a copper plate. What is the maximum kinetic energy of a photoelectron emitted in eV?

$$\upsilon = 2.5 \times 10^{15}\,\text{Hz}$$
$$\upsilon_o = 1.1 \times 10^{15}\,\text{Hz}$$

From Einstein's photoelectric equation

KE_{max}	$=$	$h\upsilon - h\upsilon_o$
	$=$	$h\,(\upsilon - \upsilon_o)$
	$=$	$(6.63 \times 10^{-34}\,\text{Js})\,(2.5 - 1.1)\,10^{-15}\,\text{Hz}$
	$=$	9.27×10^{-19} J

$1\ \text{eV} = 1.6 \times 10^{-19}$ J.

In eV
$$KE_{max} = \frac{9.27 \times 10^{-19}}{1.6 \times 10^{-19}}$$
$$= 5.79\ \text{eV}$$

8.5.1.2 Photocell

The photocell is based on the emission of photoelectrons when surfaces are irradiated by light photons.

The photocell is an evacuated cylindrical tube with two electrodes, one a semi-cylindrical surface (cathode) and the other a metallic rod (anode). The circuit diagram is shown in Figure 8.11.

Photoelectrons are produced as long as the light continues to illuminate the cathode. If interrupted, the photoelectric current stops. This effect is used in a number of devices. When intensity of light increases, the current also rises.

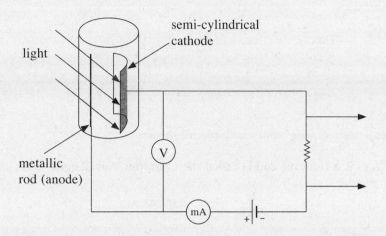

Fig. 8.11: A schematic circuit diagram for a photocell circuit.

The photocell can be employed as a switch. In a smoke alarm, the smoke reduces the intensity of light reaching the cell, thus reducing the current. This reduction in current triggers an alarm. Other such devices are garage doors and burglar alarms. If a person comes between the light and the photocell, the current becomes zero, this may turn off the garage door, switch or sensitize a warning system.

Counters also use photocells. Each time the light is interrupted, the count increases. Other applications are sound track systems, automatic systems in street lighting, etc.

8.5.2 Compton Effect

Arther H. Compton performed an experiment in 1923 which further confirmed the particle nature of light.

An x-ray beam of wavelength λ_o was directed on a slab of graphite. The x-rays scattered from the graphite crystal. The scattered x-rays were found to have a wavelength λ longer than λ_o. **The change of λ which is the amount of energy lost in moving through the graphite is proportional to the scattering angle θ.** This is known as the **Compton Effect.**

The photon is like a particle. It has both energy and momentum. When it collides with an electron, some of its energy is transferred to the electron (Figure 8.12) and the scattered photon has less energy (longer wavelength). Please recall collisions in Mechanics.

The change in wavelength of the photon is given by

$$\Delta \lambda = \lambda - \lambda_o$$

$$= \frac{h}{m_e c} (1 - \cos \theta) \qquad\qquad 8.23$$

where θ is the angle by which the photon is scattered and m_e is the mass of the electron.

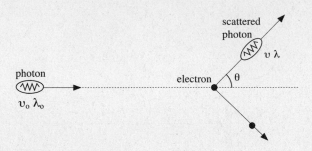

Fig. 8.12: Compton's effect showing incident and reflected photons.

The quantity $h/(m_e c)$ **is a constant and is called the Compton Wavelength.** It is equal to

$$\frac{h}{m_e c} = \mathbf{0.00243 \ nm} \qquad\qquad 8.24$$

If visible light is used in this experiment, the shift in wavelength $\Delta\lambda$, is so small that it is difficult to detect.

Example: X-rays are scattered at 50°. The incident x-rays have a wavelength of 0.21×10^{-9} m. Calculate the wavelength of the scattered x-rays.

$$\lambda_o = 0.21 \times 10^{-9} \text{ m}$$

$$\theta = 50°$$

$$h/(m_e c) = 0.00243 \times 10^{-9} \text{ m}$$

Applying Compton's equation

$$\lambda = \lambda_o + h/m_e c \ (1 - \cos \theta)$$

$$= 0.21 \times 10^{-9} \text{m} + (0.00243 \times 10^{-9}) (1 - \cos 50°)$$

$$= 0.21 \times 10^{-9} \text{m} + (0.00243 \times 10^{-9}) (0.357)$$

$$= 0.21 \times 10^{-9} \text{m} + 0.000867 \times 10^{-9} \text{m}$$

$$= 0.210867 \times 10^{-9} \text{m}$$

$$= 0.210867 \text{ nm}$$

8.5.3 Pair Production

Pair production is a direct proof that energy can be converted into matter. It has been seen that collision between a photon and an electron can transfer all the photon energy to an electron as in photoelectric effect or only partially as in Compton Effect. It is also possible for the photon to convert totally into matter and form an electron-positron pair. The positron is a positively charged electron. This phenomenon is called **Pair Production.**

The positron is the **antiparticle** of the electron. To produce a pair, energy and momentum have to be conserved. The minimum energy for a photon to convert to an electron-positron pair is equal to the rest mass energy of the two particles.

$$E_{min} \quad = \quad (h \upsilon)_{min}$$
$$= \quad 2 \ (0.51 \ \text{MeV})$$
$$= \quad 1.02 \ \text{MeV} \qquad\qquad 8.25$$

Pair production requires a heavy particle (atomic nucleus) to occur (Figure 8.13). It cannot take place in vacuum.

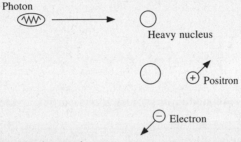

Fig. 8.13: Production of an electron-positron pair.

Test yourself: What is the largest wavelength that can form an electron-positron pair?

8.6 Pair Annihilation

Pair annihilation is the inverse of pair production. When an electron-positron pair annihilates, two photons are produced. Figure 8.14 shows the production of two photons from an electron-positron pair at rest.

A question arises: Why did not one photon form?

The electron-positron pair had zero momentum. For momentum conservation, two photons are produced which must move in opposite directions so that the momentum remains zero.

Fig. 8.14: Annihilation of an electron-positron pair.

8.7 The Wave Properties of Matter

Light (waves) have a dual nature. What about particles (matter)? Do they also exhibit wave characteristics?

In 1924, Louis de Broglie presented a revolutionary idea. He postulated that matter, like waves, must also exhibit both particle and wave properties.

The energy of a photon (equation 8.17) is

$$E \quad = \quad h\upsilon \quad = \quad hc/\lambda$$

For a particle which has no rest mass energy

$$E \quad = \quad pc$$

The momentum of a photon particle is given by

$$p \quad = \quad \frac{E}{c}$$

$$= \quad \frac{(hc/\lambda)}{c}$$

$$p \quad = \quad h/\lambda \qquad\qquad\qquad 8.26$$

The wavelength of a particle is related to its momentum (a particle quality). It is given by

$$\lambda \quad = \quad h/p$$

$$= \quad h/(mv) \qquad\qquad\qquad 8.27$$

λ **is called the de Broglie Wavelength.**

8.7.1 The Dual Nature of Light

In the previous sections, phenomena like photoelectric effect, Compton effect were discussed. They show how light interacts with matter, showing that light or electromagnetic radiations has a particle nature. Other phenomena like diffraction and interference of light prove the wave nature of light. We conclude that light has both particle and wave properties.

Light has a dual nature. It exhibits both particle and wave characteristics.

8.7.2 The Davisson-Germer Experiment

De Broglie's hypothesis that matter exhibits both wave and particle properties was verified by Davisson and Germer in the USA; and independently by G.P. Thomson in England.

Since waves exhibit particle properties (Compton and photoelectric effects), particles should exhibit wave properties like interference and diffraction.

Davisson and German beamed 54 eV electrons on a nickel crystal (Figure 8.15). The momentum associated with the electrons can be found using

$$KE \quad = \quad \tfrac{1}{2}\,mv^2$$

$$= \quad 54 \text{ e V}$$

$$= \quad 54 \times 1.6 \times 10^{-19} \text{ J}$$

$$= \quad 86.4 \times 10^{-19} \text{ J}$$

$$\text{Momentum p} \quad = \quad mv$$

$$= \quad \sqrt{(2m\ KE)}$$

$$= \quad \sqrt{[2\,(9.1\times10^{-31}\text{kg})\,(86.4\times10^{-19}\text{J})]}$$

$$= \quad 39.65 \times 10^{-25} \text{ kg.m/s}$$

The de Broglie wavelength of the electron

$$\lambda \quad = \quad \frac{h}{mv}$$

$$= \quad \frac{6.63\times10^{-34}\ \text{J.s}}{39.65\times10^{-25}\ \text{kg m/s}}$$

$$= \quad 1.67 \times 10^{-10} \text{ m}$$

According to interference theory, 1.7×10^{-10} m waves would scatter along 50°. In their experiment, Davisson and German found that the electrons were scattered at 50° (Figure 8.15).

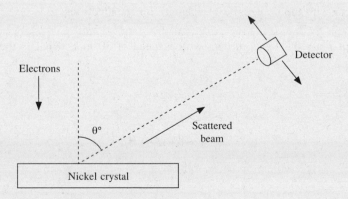

Fig. 8.15: Davisson and German experiment.

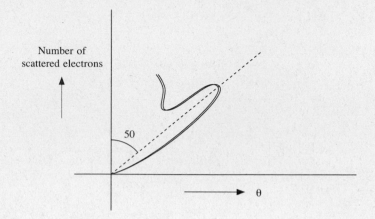

Fig. 8.16: Graph showing scattering of 54 eV electrons by Nickel crystal.

Whilst performing their experiment, Davisson and German observed distinct maxima and minima depending on the electron energy.

Since the Bragg planes of nickel crystal are not horizontal, the scattering angle was found to be 65°. Putting value in the Bragg equation for n = 1, θ = 65° and d = 0.091nm.

$$n \lambda = 2 d \sin \theta$$

$$\lambda = (2)(0.091 \text{ nm}) \sin 65°$$

$$= 0.165 \text{ nm}$$

which is in agreement with the de Broglie wavelength for the electrons.

Example: A ball of mass 0.2 kg is thrown with a speed of 30 m/s. What is the ball's de Broglie wavelength?

$$m = 0.2 \text{ kg}$$

$$v = 30 \text{ m/s}$$

$$h = 6.63 \times 10^{-34} \text{ J.s}$$

$$\lambda = \frac{h}{mv}$$

$$= \frac{6.63 \times 10^{-34} \text{ J.s}}{(0.2 \text{ kg})(30 \text{ m/s})}$$

$$= 1.105 \times 10^{-34} \text{ m}$$

This wavelength is unimaginably small.

Test yourself: What is the de Broglie wavelength of a 100 eV electron?

8.8 Application of the Wave Properties of Light—The Electron Microscope

The electron microscope has been designed on the wave properties of electrons.

The resolving power of the electron microscope is much higher than that of the light microscope. This is because electrons are accelerated to high velocities. They have shorter wavelengths. Microscopes can resolve sizes of the order of the wavelength. The wavelengths of electrons are 100 times shorter than that of the light used in optical microscopes, thus giving a high degree of resolution.

The main features of the electron microscope are the electron source, the focusing lenses, the projecting lenses and the screen.

The electrons are first accelerated. These high energy electrons have to acquire enough energy to penetrate thin slices of the specimen being examined. The electromagnetic lens focuses the electrons on to the specimen. This electron beam strikes and penetrates thin slices of sample to be viewed. The beam is scattered according to different densities within the specimen. The projector lens produces an image on the fluorescent screen which can be observed (Figure 8.17).

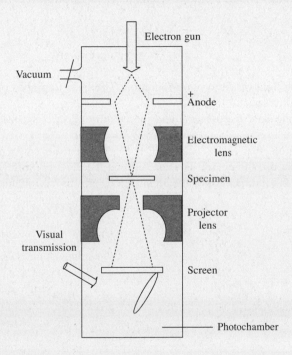

Fig. 8.17: The Electron Microscope.

More sophisticated electron microscopes are available which have additional features to obtain images which are magnified more and have higher resolution. Pictures can also be taken of the images produced using special films.

8.9 The Uncertainty Principle

Can we determine the position, speed and energy of a particle and then predict its future position and speed? Because all particles have a wave nature, it is not possible to predict future positions and speeds accurately.

8.9.1 Heisenberg's Uncertainty Principle
The Uncertainty Principle states that the product of the uncertainty in measurement of position Δx of a particle, and the uncertainty in measurement of momentum Δp cannot be less than $h/2\pi$.

In other words, both position and momentum cannot be measured with absolute precision at the same time. If position is measured exactly, the uncertainty in momentum increases and vice verse.

$$\Delta x \, \Delta p_x \quad = \quad h \qquad\qquad 8.28$$

At best the error in position is equal to one wavelength. Then equation 8.28 becomes

$$\Delta x \, \Delta p_x \quad = \quad (\lambda)\,(h/\lambda)$$

$$= \quad h \text{ is the uncertainty.}$$

An object is viewed using light or some other radiation. The light impinges on the object and returns to the eye. You then see the object.

Suppose you wish to measure the position and momentum of an electron. Light that is photons is beamed on to the electron.

The photon strikes the electron and is scattered. It can be seen through a microscope. That during the collision with the electron, the photon is scattered and the electron suffers a recoil (Figure 8.18). Because of the collision, the momentum of the electron changes by an indeterminate amount Δp_x.

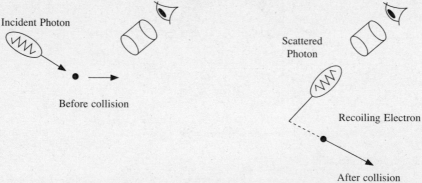

Fig. 8.18: Measuring position of an electron using a photon.

The Uncertainty Principle is so given by

$$\Delta x \, \Delta p_x \quad \geq \quad \frac{h}{2\pi} \qquad\qquad 8.29$$

For energy and time, the uncertainty principle is given by

$$\Delta E \, \Delta t \quad \geq \quad \frac{h}{2\pi}$$

Example: An electron's speed of 4×10^5 m/s is measured to an accuracy of 0.003%. Find the uncertainty in determining its position.

$$p \quad = \quad m\,v$$

$$= \quad (9.1 \times 10^{-31}\text{kg})\,(4 \times 10^5 \text{ m/s})$$

$$= \quad 3.64 \times 10^{-25} \text{ kg m/s}$$

$$\Delta p_x \quad = \quad 3.64 \times 10^{-25} \times 0.003\ \%$$

$$= \quad 3.64 \times 10^{-25} \times \frac{0.003}{100}$$

$$= \quad 10.92 \times 10^{-30} \text{ kg. m/s}$$

$$\Delta x\,\Delta p_x \quad \geq \quad \frac{h}{2\pi}$$

$$\Delta x \quad \geq \quad \frac{6.626 \times 10^{-34}\,\text{J.s}}{(2\pi)\,(10.92 \times 10^{-30}\,\text{kg m/s})}$$

$$\geq \quad 0.0966 \times 10^{-4} \text{ m}$$

$$\geq \quad 9.66 \text{ nm}$$

Example: Find the de Broglie wavelength of electrons that have been accelerated through 10 kV.

The de Broglie wavelength is given by $\lambda = \dfrac{h}{p}$

The KE of the electrons is equal to

$$KE \quad = \quad Vq$$

$$= \quad (10\text{kV})(e)$$

$$= \quad 10000 \text{ eV.}$$

In SI units

$$10000 \text{ eV} \quad = \quad 10000 \times 1.6 \times 10^{-19} \text{ J}$$

$$= \quad 1.6 \times 10^{-15} \text{ J}$$

But

$$p \quad = \quad \sqrt{(2m\,KE)}$$

$$= \quad \sqrt{(2 \times 9.1 \times 10^{-31}\text{kg} \times 1.6 \times 10^{-15}\text{J})}$$

$$= \quad \sqrt{(29.12 \times 10^{-46}\,\text{kg}^2\,\text{m}^2/\text{s}^2}$$

$$= \quad 5.39 \times 10^{-23} \text{ kgm/s}$$

And so

$$\lambda \quad = \quad \frac{h}{p}$$

$$\lambda \quad = \quad \frac{6.63 \times 10^{-34}\,\text{J s}}{5.39 \times 10^{-23}\,\text{kg m/s}}$$

$$= \quad 1.23 \times 10^{-11} \text{ m}$$

Test yourself: What is the momentum of photons of frequency 8×10^{15} Hz?

Summary

Motion is a relative quantity. There is no absolute motion.

The Postulates of the Special Theory of Relativity are:
1. The laws of physics are the same for all inertial frames (frames moving at constant velocity to each other).
2. The speed of light is a universal constant independent of the relative motion between observer and motion of the light source.

Time dilation, length contraction and variation in mass are given by

$$t = \frac{t_o}{\sqrt{(1 - v^2/c^2)}}$$

$$L = L_o \sqrt{(1 - v^2/c^2)}$$

$$m = \frac{m_o}{\sqrt{(1 - v^2/c^2)}} \quad \text{respectively.}$$

A body is moving with velocity v. It has a total energy E equal to the sum of its rest mass energy and kinetic energy, KE:

$$E = m_o c^2 + KE$$

$$KE = E - m_o c^2$$

$$= mc^2 - m_o c^2$$

$$= \Delta mc^2$$

The kinetic energy is equal to the increase in mass times (velocity)^2.

Another useful equation for the total energy is

$$E^2 = (m_o c^2)^2 + p^2 c^2$$

where KE has been written in terms of the relativistic momentum p.

If the particle is at rest (v = 0)

$$E^2 = (m_o c^2)^2$$

$$E = m_o c^2$$

and if a particle has no mass (photon)

$$E = p c$$

A black body absorbs all radiations that fall on it.

The characteristics of blackbody radiation cannot be explained using classical concepts. The peak value of a blackbody radiation curve is given by Wien's Displacement Law:

$$\lambda_{max} T = 2.898 \times 10^{-3} \text{ m.K}$$

where λ_{max} is the wavelength at which the curve peaks and T is the absolute temperature of the object emitting the radiation.

Stefan and Boltzmann found that the amount of radiation emitted by a body is proportional to the body's surface area and the absolute temperature

$$E \quad \alpha \quad AT^4$$

$$= \quad \sigma AT^4$$

where $\sigma = 5.67 \times 10^{-8}$ W/(m^{2}K^4) is the constant of proportionality and is known as the Stefan-Boltzmann Constant.

Planck first introduced the quantum concept when he assumed that the subatomic oscillators responsible for blackbody radiation could have only discrete amounts of energy given by

$$E_n = nh\upsilon \qquad\qquad n = 1, 2, 3,\ldots\ldots\ldots$$

where n is a positive integer and υ is the frequency of vibration of the resonator.

Electromagnetic radiation and matter interact by three processes:
1. The photoelectric effect is a process whereby electrons are ejected from a metal surface when light is incident on that surface. Einstein provided a successful

explanation of this effect by extending Plank's quantum hypothesis to electromagnetic waves. In this model, light is viewed as a stream of particles called photons, each with energy $E = h\upsilon$. The maximum kinetic energy of the ejected photoelectrons is

$$KE_{max} = h\upsilon - \Phi_o$$

Φ_o is the work function of the metal.

2. X-rays are scattered by electrons in matter. In such a scattering event, a shift in wavelength is observed for the scattered x-rays. This phenomenon is known as the Compton Shift. Conservation of momentum and energy applied to photon-electron collision yields the following expression for the shift in wavelength of the scattered x-rays:

$$\Delta\lambda = \lambda - \lambda_o = \frac{h}{m_e c}(1 - \cos\theta)$$

where m_e is the mass of the electron, c is the speed of light, and θ is the scattering angle.

3. High energy photons like γ rays interact with matter to form an electron-positron pair. The minimum energy of the γ ray photon must be equal to the rest energy of the pair. It is equal to

$$E_{min} \geq (h\upsilon)_{min}$$
$$\geq 1.02 \text{ MeV}$$

De Broglie proposed that all matter has both a particle and a wave nature. The de Broglie wavelength of any particle of mass m and speed v is

$$\lambda = h/p = h/mv$$

De Broglie also proposed that the frequencies of the waves associated with particles obey the Einstein relationship $E = h\upsilon$.

According to Heisenberg's Uncertainty Principle, it is impossible to measure simultaneously the exact position and exact momentum of a particle. If Δx is the uncertainty in the measured position and Δp_x the uncertainty in the momentum, the product $\Delta x\, \Delta p_x$ is given by

$$\Delta x\, \Delta p_x \geq \frac{h}{2\pi}$$

and

$$\Delta E\, \Delta t \geq \frac{h}{2\pi}$$

where ΔE is the uncertainty in the energy of the particle and Δt is the uncertainty in the time it takes to measure the energy.

Questions

1. Are blackbodies black in colour?
2. We are unable to see objects in the dark although all objects radiate energy. Why?
3. A photon is scattered by the Compton effect. Can its wavelength become shorter?
4. A certain metal emits photoelectrons. Will a different metal also emit photoelectrons under the same conditions? Explain.
5. Which has more energy, a photon of ultraviolet radiation or a photon of green light?
6. Which photon has the greater energy, one of blue light or one of an X-ray?
7. How does energy of a quantum of light vary with its wavelength?
8. When a photon of radiation falls on the surface of a photoelectric cell, it may completely disappear. What happens to its energy?
9. Why are no photoelectrons emitted from the metal when the frequency of light is below the threshold value?
10. When an electron and positron annihilate, why are two photons produced?
11. Why do we not observe the wave properties of ordinary objects such as books and pens?

12. How does the de Broglie wavelength of a particle relate to its linear momentum?

13. A car and a proton have the same wavelength. Which has greater kinetic energy?

14. The uncertainty principle is applicable to all objects, but it is significant to only small objects like electrons and protons. Why?

15. What other information would be needed to determine the rest mass of moving particle if its de Broglie wavelength is known?

16. Express the speed of a photoelectron ejected from a surface in terms of frequency of light incident on the surface, Planck's constant, the work function of the surface and the mass of the electron.

17. Solve the de Broglie equation so as to express the speed of a particle in terms of its mass, wavelength and Planck's constant.

18. Would the heart beat of an astronaut moving at high speed in a spacecraft appear to be faster or slower when observed from Earth?

Problems

1. Charged particles moving with a speed of 0.97c disintegrate in 25 μs. What would their life spans be if they were at rest?

2. A spaceship measures 150 m in length when at rest with respect to an observer. If it moves with a velocity of 0.95 c, what length will the observer measure?

3. The sun's surface temperature is 6000 K. If the sun is considered as a black body, what would be the value of the maximum intensity wavelength?

4. A kiln has a temperature of 1500 K. What is the frequency of the radiation emitted at the maximum intensity if the kiln is assumed to be a black body?

5. At a steady temperature of 1200 K, a black body radiates all energy E supplied to it. If the energy supplied to it was decreased to half E i.e. E/2, what would be its steady temperature now?

6. Radiation of wavelength 250 nm is incident on a metal. The maximum kinetic energy of the photoelectrons emitted is found to be 9×10^{-20} J. What is the work function of the metal?

7. Incident light on a sodium surface is of wavelength 0.3 μm. The work function is 2.46 eV. Find the maximum kinetic energy of the photoelectrons produced and the cut off wavelength for sodium.

8. When light of 3.75×10^{-7} m falls on a certain material, a potential difference of 0.86 V is required to stop the electrons from reaching the anode. What is the work function of the material?

9. What is the lowest frequency light that will cause the emission of photoelectrons from a surface whose nature is such that 1.9 eV are required to eject an electron?

10. An x-ray photon with a frequency of 10×10^{18} Hz bombards a substance and ejects a photoelectron with a kinetic energy of 4.9×10^{-16} J. What are the wavelengths of the incident and emergent photons?

11. Calculate the scattering angle for x-rays scattered by a carbon target. The wavelength shift is 1.5×10^{-3} nm.

12. A 0.15 nm x-ray photon collides with an electron at rest. The scattered photon moves at an angle of 80°. Determine the Compton shift.

13. Calculate the energy and momentum of a photon of wavelength 750 nm?

14. The eye can detect very small energies of the order of 10^{-18} J. How many photons of energy of wavelength 6×10^{-7} m is this energy equal to?

15. What is the mass of a photon of wavelength 800 nm?

16. A sodium lamp has a power output of 1000 W. Using 589.3 nm as the average wavelength of this source, find the number of photons emitted per second.

17. How many photons would be emitted per second by a yellow light lamp radiating at a power of 15 W? (Frequency of yellow light is 5×10^{14} Hz).

18. The most prominent colour in sunlight is yellow light. Its frequency is 6×10^{14} Hz. What is the energy of a photon that has this frequency?

19. What are the frequency and the energy of a quantum of red light with a wavelength of 6.4×10^{-7} m?

20. What is the de Broglie wavelength of a car with a 1000 kg mass and speed of 90 km/hr?

21. What is the mass associated with a 80 μm photon?

22. What is the de Broglie wavelength of a neutron with mass of 1.67×10^{-27} kg and a speed of 10^5 m/s?

23. The wavelength of an electron is 3×10^{-11} m. Through what potential difference did the electron fall and what was its speed?

24. The uncertainty in the position of a proton is 10 nm. What is the uncertainty in its velocity?

25. If the uncertainty in the position of a baseball as it leaves the pitcher's hand is 6 cm, what is the uncertainty in its momentum?

9 Atomic Spectra

OBJECTIVES

After studyig this chapter, the student should be able:
- to understand the hydrogen emission spectrum.
- to understand Bohr theory of the hydrogen atom.
- to understand de Broglie's explanation of the hydrogen atom.
- to understand the uncertainty principle.
- to understand the application of the uncertainty principle to show that electrons cannot exist in the nucleus of an atom.
- to understand how x-rays are produced as well as their properties and uses.
- to distinguish between continuous and characteristic x-ray spectra.
- to understand lasers and.
- to understand how He-Ne laser is produced and uses of lasers.

9.1 Introduction

All matter is made up of atoms. The atom and its structure are important to understand the properties of different materials. Bohr's model of the hydrogen atom and later the quantum mechanical description has helped in explaining the emission of electromagnetic radiation including light by the atom as well as in the development of technology for the production of x-rays and lasers. Both lasers and x-rays have important applications in today's world.

9.2 Atomic Spectra

When gas or vapour at pressures less than that of the atmospheric pressure is excited by an electric current, light of specific (discrete) wavelengths are emitted (Figure 9.1). These wavelengths characterize the element whose vapour was excited.

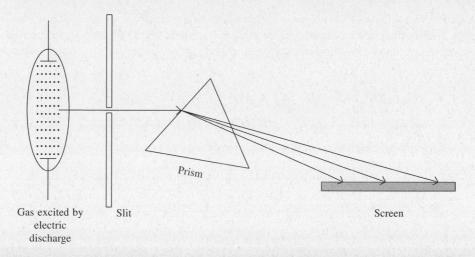

Fig. 9.1: Separation of the line spectra of the excited gas vapour by a prism.

These atomic spectra are called **line spectra** because of their appearance. The line spectra of hydrogen, helium and mercury are shown in Figure 9.2.

Fig. 9.2: Spectra of hydrogen, helium and mercury in visible light region.

These line spectra have helped in identifying elements because each element has its own unique line spectrum.

9.2.1 Atomic Spectra of Hydrogen

Line spectra are also produced in the ultraviolet and infrared regions. The line spectrum (pattern of lines) obtained for hydrogen atom have been studied in great detail. A number of series (groupings) of lines were observed in the visible and non-visible regions of the electromagnetic spectrum for hydrogen. Johann J. Balmer in the late nineteenth century derived equations for the observed spectral lines. Similar equations were also determined for other groups of lines in the hydrogen spectrum empirically. The equations are:

Lyman Series $1/\lambda = R[1/1^2 - 1/n^2]$ n = 2, 3, 4, ...   9.1

Balmer Series $1/\lambda = R[1/2^2 - 1/n^2]$ n = 3, 4, 5, ...

Paschen Series $1/\lambda = R[1/3^2 - 1/n^2]$ n = 4, 5, 6, ...

In the above equations, R is the **Rydberg's Constant. Its value is R = 1.097 × 10^7 m^{-1}**.

It was observed that the lines in each group, crowd near the short wavelengths (Figure 9.3).

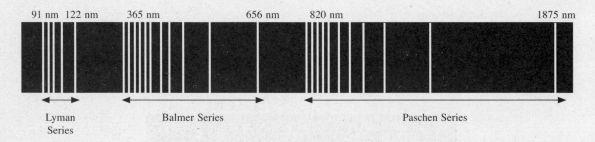

Fig. 9.3: Line spectrum of hydrogen.

In the hydrogen spectrum, only the Balmer Series are in the visible range.

9.3 Classical Model of the Hydrogen Atom

After Rutherford's famous experiment, a new picture of the atom emerged. The atom has a small, massive, positively charged nucleus around which electrons revolve in stable orbits.

The hydrogen atom which is the simplest of all atoms has a nucleus of one proton and orbiting around this nucleus is an electron. The amount of work needed to remove the electron from the atom is 13.6 eV. This energy means that the electron's orbit has a radius:

$$r = 5.29 \times 10^{-11} \text{ m} \qquad\qquad 9.2$$

and its velocity is: $v = 2.2 \times 10^6$ m/s.   9.3

The proton is 1836 times more massive than the electron and so can be considered stationary. The electromagnetic theory predicts that accelerated charges radiate energy. Thus, the electron would continue to radiate energy as it revolves around the nucleus. Its orbit would become smaller and smaller till it hits the nucleus (Figure 9.4).

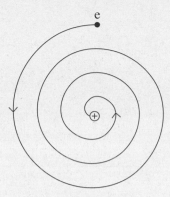

Fig. 9.4: Radiating e spirals into the nucleus.

The classical laws of physics cannot explain the stationary orbits of the hydrogen atom. Quantum mechanical concepts which explained the photoelectric and Compton's effects might have the answers to this dilemma.

Bohr gave an explanation for the stationary orbits. He postulated:

1. In the hydrogen atom, only certain values of the total energy for the electron were allowed. These allowed energies are equivalent to specific orbits. When the electron revolves in these orbits, it does not radiate energy.

2. An electron can orbit around the nucleus only if the orbit circumference is equal to an integral number of its wavelengths.

$$n \lambda = 2 \pi r_n \qquad n = 1, 2, 3, \ldots \qquad\qquad 9.4$$

r_n is the radius of the orbit which has an integer number n of wavelengths. The number n is called the quantum number of the orbit.

3. The electron radiates energy when it jumps from a higher energy level to a lower energy level.

9.4 Bohr's Theory of the Hydrogen Atom

Bohr in 1913 postulated his classical model of the hydrogen atom incorporating the quantum ideas of Max Planck and Einstein. According to his model, the electron can only orbit in certain allowed orbits around the nucleus. It cannot be in an orbit in-between the allowed orbits. These allowed orbits are called **stable energy levels** or **stationary states**. In these orbits, no energy is radiated. To move to a higher orbit, the electron must acquire energy equal to the energy difference of the two orbits. The **ground state** is the lowest preferred and most stable energy level.

The higher energy levels are called **excited levels.** If the electron acquires enough energy to leave the atom, the atom is ionized and the minimum energy needed to leave the atom is called **ionization energy.**

Photons are emitted when the electron jumps from a higher to a lower orbit (Figure 9.5)

The energy change is related to the frequency of the light emitted as

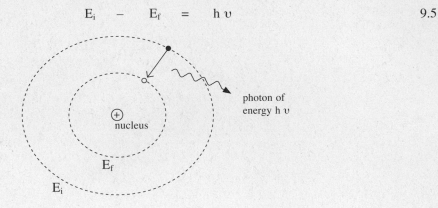

$$E_i \quad - \quad E_f \quad = \quad h\, \upsilon \qquad \qquad 9.5$$

Fig. 9.5: Emission of photon when an electron jumps from a higher to a lower orbit.

The de Broglie wavelength λ is

$$\lambda \quad = \quad \frac{h}{mv}$$

$$\lambda \quad = \quad \frac{6.63 \times 10^{-34} \text{ J.s}}{9.1 \times 10^{-31} \text{ kg} \times 2.2 \times 10^{6} \text{ m/s}}$$

$$= \quad 3.3 \times 10^{-10} \text{ m}$$

This result exactly matches the circumference of the electron's orbit.

$$2\,\pi\,r = \ 2\,\pi \times 5.3 \times 10^{-11} \text{ m}$$

$$= \ 3.3 \times 10^{-10} \text{ m}$$

The orbit of the electron is equal to one electron wave. When a vibrating wire loop is examined, it is found that when an integral number of wavelengths match the loop circumference, vibrations continue indefinitely if there are no resistive losses (Figure 9.6).

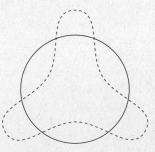

Fig. 9.6: A vibrating loop equal to 3 wavelengths.

Bohr used the idea of such waves to make certain postulates.

9.4.1 Bohr Postulates

Bohr's postulates are:

1. The electron can orbit around the nucleus in particular (discrete) orbits only. In these orbits, it does not radiate energy.
2. The allowed orbits r_n are those for which the orbital angular momentum is an integral multiple of $h/2\pi$.

$$m \, v \, r_n \; = \; \frac{nh}{2\pi} \qquad\qquad 9.6$$

where n is an integer (n = 1, 2, 3, ...), h is Planck's constant and m and v, the mass and velocity of the electron respectively.

3. When an electron moves from a higher energy level E_i to a lower energy level E_f, photons are emitted.

$$E_i \; - \; E_f \; = \; h\,\upsilon$$

where υ is the frequency of the photon and n is an integer.

De Broglie using the idea of matter waves proved Bohr's assumption. Let the circumference of the electron's orbit of radius r be $2\pi r$. For stationary waves to be maintained, the orbit must have an integral number of waves:

$$n \, \lambda \; = \; 2 \, \pi \, r$$

$$\lambda \; = \; \frac{2 \, \pi \, r}{n} \qquad\qquad 9.7$$

Equating de Broglie's relationship for λ

$$\lambda \; = \; \frac{h}{mv}$$

with equation 9.7 and rearranging the terms, de Broglie proved Bohr's second assumption

$$\mathbf{m \, v \, r} \; = \; \frac{n \, h}{2 \, \pi}$$

9.4.2 Energies and Radii of Bohr's Model

For an electron to revolve in a circular orbit, there has to be a centripetal force. The centripetal force is provided by the Coulomb's force between the nuclear charge Ze and electron's charge e

$$\text{Centripetal force} \; = \; \text{Coulomb's force}$$

$$\frac{mv^2}{r} \; = \; \frac{k\,(Ze)\,e}{r^2}$$

$$mv^2 \; = \; \frac{k\,Z\,e^2}{r} \qquad\qquad 9.8$$

The electron moving around its orbit has both kinetic and potential energies. The kinetic energy is $\frac{1}{2}mv^2$ and the potential energy is given by

$$
\begin{aligned}
\text{PE} &= \text{e V} \\
&= (-\text{e}) \left(k\frac{Ze}{r} \right)
\end{aligned}
$$

as the potential at distance r from the nuclear charge (Ze) is k(Ze)/r. From equation 9.8, the kinetic energy is equal to $KE = \frac{1}{2} m v^2 = (kZe^2)/r$.

The total energy is the sum of kinetic and potential energies.

$$
\begin{aligned}
\text{E} &= \text{KE} + \text{PE} \\
&= \frac{1}{2}\frac{kZe^2}{r} - \frac{kZe^2}{r} \\
&= -\frac{kZe^2}{2r}
\end{aligned}
$$
9.9

Using Bohr's second postulate

$$ m\, v_n\, r_n = \frac{n h}{2\pi} \qquad n = 1, 2, 3, \ldots. $$

$$ v_n = \frac{n h}{2\pi m\, r_n} $$

Inserting the value of v in equation 9.8 results in:

$$ m v^2 = \frac{kZe^2}{r} $$

$$ \frac{m n^2 h^2}{(2\pi)^2 m^2 r_n{}^2} = \frac{kZe^2}{r} $$

$$ \frac{n^2 h^2}{4\pi^2 k Z e^2 m} = r_n \qquad n = 1, 2, 3, \ldots. $$
9.10

Putting in values of m and e, h and k, gives the radii for the Bohr orbits as

$$ r_n = (5.29 \times 10^{-11}) \frac{n^2}{Z} \qquad n = 1, 2, 3, \ldots. $$
9.11

The radii are quantized.

For hydrogen $\quad Z = 1$ and so

$$ r_n = (5.29 \times 10^{-11}) n^2 $$
9.12

The ground state orbit (n = 1) radius for hydrogen is

$$ r_1 = 5.29 \times 10^{-11}\ m $$
9.13

This is known as the **Bohr radius.**

  The total energy can be determined using equation 9.9:

$$E_n = -\left[\frac{2\pi^2 \, m \, k^2 \, e^4}{h^2}\right] \frac{Z^2}{n^2} \qquad n = 1, 2, 3, \ldots \qquad 9.14$$

Putting values for m, k, e and h gives

$$E_n = \frac{-(2.18 \times 10^{-18}]\, Z^2}{n^2} \; \text{joules}$$

Which in electron-volts (eV) by dividing with the charge of an electron i.e. 1.6×10^{-19} becomes

$$E_n = -(13.6 \text{ eV}) \frac{Z^2}{n^2} \qquad n = 1, 2, 3, \ldots \qquad 9.15$$

For **hydrogen Z = 1, the total energy is:**

$$E_n = \frac{-13.6}{n^2} \text{ eV} \qquad n = 1, 2, 3, \ldots \qquad 9.16$$

and $\qquad\qquad E_1 = -13.6 \text{ eV} \qquad\qquad$ as $n = 1$ $\qquad\qquad$ 9.17

The energy levels for stationary orbits of hydrogen are shown in Figure 9.7.

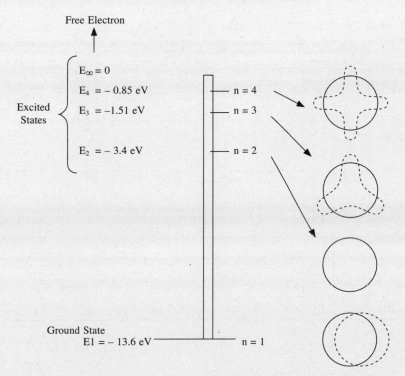

Fig. 9.7: Energy levels of the hydrogen atom.

9.5 Emission Spectra of Hydrogen Atom

The line spectra for hydrogen can be determined from the energy equation 9.5 and c = υ λ

$$h\upsilon = E_i - E_f$$

$$\frac{1}{\lambda} = \frac{2\pi^2 m k^2 e^4 Z^2}{h^3 c}\left[\frac{1}{n_f^2} - \frac{1}{n_i^2}\right] \qquad 9.18$$

where n_i = 1, 2, 3, …… and n_f = 1, 2, 3,………. When values of m, k, e, h and c are put in equation

$$\frac{1}{\lambda} = (1.097 \times 10^7 \text{ m}^{-1})\, Z^2 \left[\frac{1}{n_f^2} - \frac{1}{n_i^2}\right]$$

$$= R Z^2 \left[\frac{1}{n_f^2} - \frac{1}{n_i^2}\right]$$

Here n_i is greater than n_f and 1.097×10^7 m^{-1} is equal to the Rydberg's constant R. Thus, Bohr's theory was able to experimentally verify the value of the Rydberg's constant.

When Z = 1 (hydrogen) and n_f = 1, transitions from higher levels n_i = 2, 3, 4,…. to n_f = 1 are known as the Lyman Series; transitions to n_f = 2 from higher energy levels are the Balmer Series and the Paschen Series are transitions from n_i = 4, 5, 6, …… to n_f = 3.

Figure 9.8 shows the Lyman, Balmer and Paschen series of the line spectrum of the hydrogen atom.

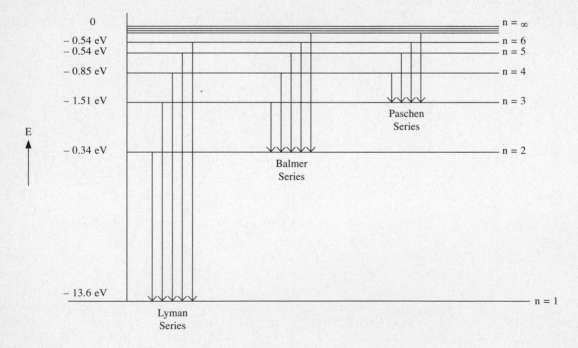

Fig. 9.8: Lyman, Balmer and Paschen series of hydrogen.

Example: Determine the longest wavelength of the Lyman series. Lyman series are transitions to $n_f = 1$ from higher levels. The longest wavelength will be from $n_i = 2$ to $n_f = 1$.

$$R = 1.097 \times 10^7 \text{ m}^{-1}$$

$$n_i = 2$$

$$n_f = 1$$

Using equation 9.1

$$\frac{1}{\lambda} = (1.097 \times 10^7 \text{ m}^{-1})\left[\frac{1}{1^2} - \frac{1}{2^2}\right]$$

$$= (1.097 \times 10^7 \text{ m}^{-1})\frac{3}{4}$$

$$= 0.82275 \times 10^7 \text{ m}^{-1}$$

$$\lambda = 1.215 \times 10^{-7} \text{ m}$$

9.5.1 Absorption Spectrum of Hydrogen

If hydrogen gas is irradiated with white light, some wavelengths disappear. When the continuous spectrum of white light is observed, a number of dark lines are seen (Figure 9.9). This is the absorption spectrum of hydrogen. These dark lines correspond to the hydrogen atom spectral lines discussed in the last sections.

The energies absorbed by the hydrogen atoms must be equal to the energy difference between two energy levels. This energy is absorbed by the electron to jump from a lower to a higher level.

Fig. 9.9: Absorption spectra of hydrogen.

9.6 X-rays — Inner Shell Transitions and Characteristic X-rays

The photoelectric effect showed that a photon could transfer its energy to electrons. Is the reverse also true? Can an energetic electron convert some or all of its kinetic energy into a photon?

This effect was discovered by Roentgen in 1895. He found that accelerated electrons, when stopped by a **heavy** metal target, produce a very penetrating radiation. This radiation was called **x-rays.**

To remove electrons from shells closer to the nucleus in heavy elements, high values of energies are needed. In the laboratory, x-rays are produced when electrons are accelerated through a very high potential difference. These very fast moving electrons are then stopped by a heavy metal target like platinum. In Figure 9.10, the target anode A is molybdenum, G is a glass tube that has been evacuated, and C is the cathode with a filament to produce electrons. A very high potential difference is applied between A and C.

Electrons are emitted when the current in the filament circuit heats the filament. These electrons are called thermions. If the accelerating voltage is V, the electrons acquire kinetic energies KE equal to

$$KE \quad = \quad \tfrac{1}{2}\,mv^2 \quad = \quad eV \qquad\qquad 9.20$$

Where e is the electronic charge, m the mass and v the speed of the electrons. As the electrons stop, their energy is released in the form of x-rays.

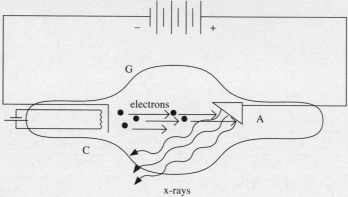

Fig. 9.10: An x-ray tube.

The graph in Figure 9.11 shows the intensity of the waves and the wavelength that is produced.

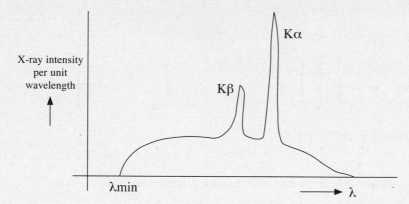

Fig. 9.11: X-ray spectrum of molybdenum at 35 kV accelerating potential.

The graph is a smooth continuous spectrum with sharp peaks K_α and K_β superimposed on it. **The continuous spectrum is called Bremsstrahlung** (braking radiation) and the sharp peaks are the **characteristic x-rays.**

9.6.1 Continuous X-ray Spectrum

The fast moving electrons are slowed down (decelerated) when they hit the target atoms. The electron may stop after a single impact or in a couple of impacts. The energy lost in collisions produces x-rays of different intensities. These x-rays are responsible for the **continuous spectrum.**

If an electron gives up all its kinetic energy in one collision, the x-ray emitted will have the shortest wavelength λ_{min}. This would correspond to kinetic energy KE = $h\upsilon_{max}$. On the other hand, an electron which has kinetic energy k_o collides with an atom of the target. It loses energy Δk which will be radiated as x-ray. The scattered electron still has energy $k_o - \Delta k$ left (Figure 9.12). This electron will again collide with other atoms and lose more energy in the form of x-rays until the electron loses all its kinetic energy.

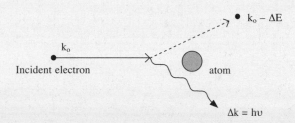

Fig. 9.12: An incident electron of energy k_o, scattered electron with energy $k_o - \Delta k$ and x-ray of energy Δk.

9.6.2 Characteristic X-rays

Like light, x-rays are also produced by electronic transitions. As x-rays have shorter wavelengths and higher frequencies, they are produced by **heavy** atoms in which energy levels have higher values than those of lighter elements. In atoms with more than one electron, electrons having the same value of quantum number n, are in the same **shell.** For instance, n = 1 is called K shell, n = 2 is called L shell and so on. These shells are further divided into sub-shells with different quantum numbers like l, m_l.

When electrons are accelerated to produce x-rays, a few electrons might gain enough energy to penetrate the atom and remove an electron from the lower orbits. Electrons from the higher orbits fall to the empty space and release their excess energy in the form of x-rays.

For example, the second sharp peak in the Figure 9.11 is due to an electron transition from n = 2 shell falling into a vacancy of n =1 shell. It is known as K_α line. The line K_β is produced when a transition takes place from n = 3 to n = 1. These are the **characteristic x-rays spectral lines** (Figure 9.13). The frequency $\upsilon_{K\alpha}$ of the x-ray K_α spectral line can be obtained by applying Planck's equation

$$E_L \quad - \quad E_K \quad = \quad h \, \upsilon_{K\alpha}$$

and $\upsilon_{K\beta}$ for K_β

$$E_M \quad - \quad E_K \quad = \quad h \, \upsilon_{K\beta} \qquad\qquad 9.21$$

where E's are energies at levels K, L and M and $\upsilon_{K\alpha}$ and $\upsilon_{K\beta}$ the frequencies of the characteristic x-rays.

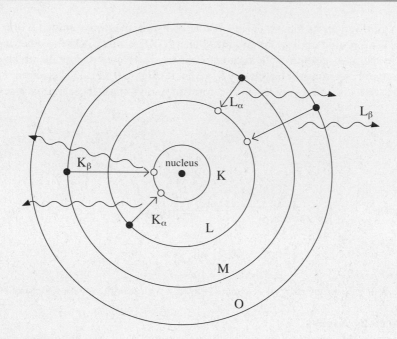

Fig. 9.13: The transitions for K_α, K_β, L_α and L_β x-ray spectral peaks.

9.6.3 Properties of X-rays

1. X-rays are electromagnetic waves. They have short wavelengths between $10^{-9} - 10^{-12}$ m.
2. They are not deflected by electric or magnetic fields as they have no charge. They move in straight lines.
3. They are very penetrating. The faster the speed of the electrons, the more penetrating are the x-rays produced. Less x-rays are absorbed by fleshy tissue as compared to denser material like bones. This is because bones are made up of more heavy (higher atomic number) elements.
4. X-rays ionize the material they pass through.
5. X-rays affect photographic plates and make phosphorescent substances glow.
6. X-rays exhibit interference and diffraction.

9.6.4 Uses of X-rays

1. X-rays are used to photograph dense body parts. They pass readily through soft tissues but are stopped by the denser bones. Thus, bones can be photographed using soft x-rays (x-rays with longer wavelengths). Low radiation doses are used for taking these x-ray photographs.
2. X-rays are used in medical therapy. Hard (shorter wavelength) x-ray beams are focused on body tumours. The x-rays destroy cancerous cells. Growing and dividing cells are more sensitive to radiation.
3. X-rays are used in industry to detect flaws in welded joints.
4. X-rays diffraction patterns are used to determine crystal structures.

9.6.5 Computerized Axial Tomography (CT scan)

The CT scanner takes a number of x-rays of a patient from all angles. The scattered x-rays are fed to a computer which combines them to give a cross-sectional view of the portion of the body which has been scanned (Figure 9.14). As mentioned earlier, x-rays are absorbed more in denser materials. The density changes from those of a normal brain CT scan can be read from the three dimensional picture obtained. By this method, the problem (tumour) is not only detected but its exact position is also determined.

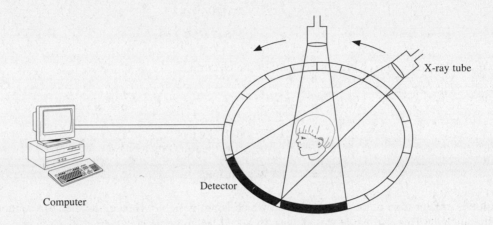

X-ray tube

Detector

Computer

Fig. 9.14: A CT scanner.

9.6.6 Biological Effects of X-rays
X-rays, when focused on tumours, can destroy them.

On the other hand, x-rays may cause damage to body tissues and cells. Cell mutations caused by x-rays may be passed on to the next generation. They may also cause cancer by producing free radicals. These radicals are produced when these radiations break up bonds between molecules, thus releasing radicals. The radicals interfere with cell reactions, cause DNA mutations and can also cause cancer.

It is advisable not to go for x-rays unnecessarily. X-rays should only be taken when they are important for diagnosis and treatment.

9.7 Uncertainty within the Atom

According to Heisenberg's Uncertainty Principle (Chapter 8), there is a fundamental limitation to measuring accurately the position and momentum of a particle simultaneously. This uncertainty in measurement is given by

$$\Delta x \, \Delta p_x \quad \geq \quad \frac{h}{2\pi}$$

Although in the macro world, this uncertainty in measurement is insignificant, it becomes very significant in the atomic and subatomic situation.

We can use the uncertainty principle to determine whether electrons are present in nuclei. The uncertainty in the measurement of position of the electron in the nucleus is of nuclear dimensions. The nucleus has a radius of 10^{-15} m.

The uncertainty in momentum Δp_x is

$$\Delta p_x \geq \frac{h}{2\pi(\Delta x)}$$

$$\geq \frac{6.63 \times 10^{-34} \text{ J.s}}{(2\pi)(10^{-15} \text{ m})}$$

$$\geq 1.055 \times 10^{-19} \text{ kg.m/s}$$

Using Δp_x to determine the change in velocity of the electron:

$$\Delta v = \frac{\Delta p}{m} \qquad\qquad p = m\,v$$

$$= \frac{1.055 \times 10^{-19} \text{ kg.m/s}}{9.1 \times 10^{-31} \text{ kg}}$$

$$= 1.159 \times 10^{11} \text{ m/s !!}$$

This value is greater than c, the speed of light, which is not possible. Hence, the electron cannot exist inside the nucleus. To exist inside the nucleus, its speed has to be greater than c which is impossible.

The same calculation can be made to check the existence of the electron within the atom. The atom has a radius of 10^{-10} m

$$\Delta v = \frac{6.63 \times 10^{-34} \text{ J.s}}{2\pi\,(9.1 \times 10^{-31} \text{ kg})\,(10^{-10} \text{ m})}$$

$$= \frac{1.055 \times 10^{-24} \text{ J.s}}{(9.1 \times 10^{-31} \text{ kg})}$$

$$= 0.1159 \times 10^7 \text{ m/s}$$

$$= 1.159 \times 10^6 \text{ m/s}$$

$$< c$$

This value is less than the velocity of light, hence the electron can exist within the atom.

The uncertainty principle is useful in determining whether an electron can exist in the confines of a nucleus or an atom.

9.8 The Laser

Lasers are the most important invention of the last century. They have proved their worth in numerous applications — eye and cauterization surgeries, holograms (3–D images), fibre optics, etc.

The word LASER is an acronym of **L**ight **A**mplification by **S**timulated **E**mission of **R**adiation. As the name suggests, lasers are produced by stimulated emission; and the emitted light is amplified.

9.8.1 Spontaneous Emission

Keeping in mind the quantum mechanical picture of the atom, when an electron moves from a higher to a lower energy level, radiation is emitted. In the thousands of atoms that form a gas, a number of electrons may spontaneously move from one energy level to another, producing light photons in different directions (Figure 9.15).

9.8.2 Stimulated Emission

If more such electrons are **forced** to jump to lower orbits, more concerted light could be produced. In stimulated emission, the gas is bombarded with photons which trigger electrons into changing their energy levels. In this way, instead of one, two photons are emitted for each electron jump. Care has to be taken that the incoming photons have precisely the energy which is the difference of energy between two levels. This energy difference is given by

$$\Delta E = E_{higher\ level} - E_{lower\ level}$$

$$h\upsilon = E_i - E_f$$

$$\upsilon = \frac{E_i - E_f}{h} \qquad\qquad 9.23$$

The bombarding photon must have a frequency υ given by equation 9.23.

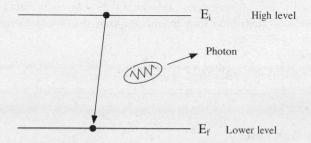

Fig. 9.15: Spontaneous emission — Photons emitted in any direction.

Stimulated emission has the following characteristics:
1) The emitted photon and the incoming photon have the same direction.
2) The two photons are coherent. They have the same phase.
3) As already mentioned, one photon goes in and two come out for each transition, i.e. amplification occurs.

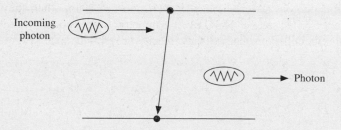

Fig. 9.16: Stimulated emission. Photon emitted in direction of incoming photon.

9.8.3 Population Inversion

Under normal situation, the lower energy levels are filled whilst the higher levels have fewer electrons and are not completely filled. If the electrons are excited by some strong sources of light, electrons in the lower energy levels would gain energy, become excited and jump to higher orbits. A situation would now exist where higher energy levels would fill up and there would be more unfilled lower energy levels. This condition when more electrons are excited to higher energy levels is called **population inversion** (Figure 9.17).

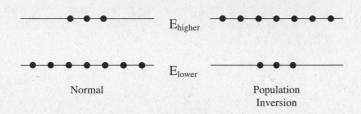

Fig. 9.17: Normal population distribution and population inversion.

The most important aspect of laser production is that many atoms are present in the excited state for longer times (10^{-3} s) rather than in the normal state, where they remain for shorter times (10^{-8} s). These states in which the electrons remain for a longer time are known as **metastable states.**

The laser beam is characterized by:
1) It is monochromatic.
2) It is coherent – all waves are in phase.
3) It does not diverge.
4) It is extremely intense.

9.8.4 Helium – Neon Gas Laser

Helium-Neon lasers developed in 1961 are the most common type of lasers. A glass tube has two parallel mirrors at the ends. One mirror is partially transparent. The tube is filled with helium and neon at low pressure in the ratio of 10:1 (Figure 9.19).

The distance between the mirrors is an integral multiple of a half-wavelength of laser light. When a high voltage discharge is passed through the helium-neon mixture, electrons of both elements absorb energies and jump to their metastable states of 20.61 eV and 20.66 eV from their ground states for helium and neon respectively. Excited helium atoms transfer their excitation and kinetic energies during collisions with neon atoms to neon atoms. Population inversion occurs for Neon atoms.

Neon electrons emit photons (laser) light of wavelength 632.8 nm when they move from excited state 20.66 eV to a less excited state at 18.70 eV. This photon triggers other stimulated emissions. The electrons lose energy in collisions before they reach the ground state. Figure 9.18 shows these transitions.

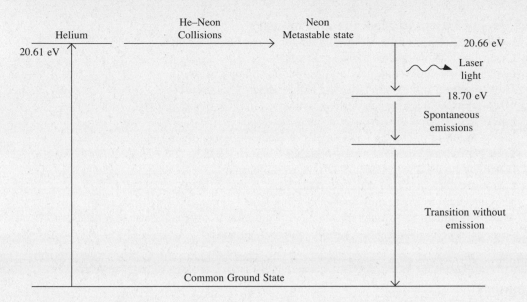

Fig. 9.18: The helium – neon laser.

The helium-neon laser operates continuously and a red laser light is produced. Population inversion is achieved at 20.66 eV in neon gas atoms. It is important to realize that a very small fraction of the atoms present are lasing at a particular moment.

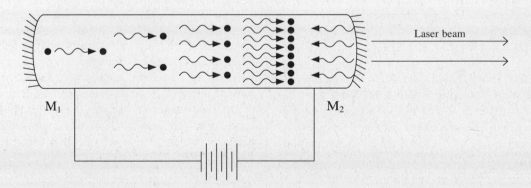

Fig. 9.19: Helium – Neon Laser.

The laser light photons are reflected at the mirrors M_1 and M_2. In this way, the discharge tube is accumulating more and more photons. The laser light leaves the discharge tube through the transparent portion of the partially reflecting mirror M_2 (Figure 9.19). The laser beam is intense and extremely narrow.

9.8.5 Uses of Lasers in Medicine and Industry

1. Lasers are used in transmission of sound data over optical fibres.
2. Lasers are used in nuclear fusion research.
3. Lasers have countless military and astronomical applications.
4. They are used in the manufacturing and reading of compact discs.
5. They are used in surgery to destroy tumours, cauterize blood vessels.
6. Lasers can be used in surveying to measure distances.
7. Lasers are used in cutting many layers of cloth in garment factories.
8. They can be used to weld metals together.
9. Laser can be used to generate 3-D images (holograms).
10. Lasers are used in fixing detached retinas and in corrective eye surgery.
11. They are used to break up kidney and gall bladder stones into small pieces.
12. A laser beam can be grinded through an optical fiber to open up clogged arteries.
13. Lasers can be used in cancer therapy to destroy tumours.

Summary

The hydrogen spectrum series are given by:

Lyman Series
$$1/\lambda = R[1/1^2 - 1/n^2] \quad n = 2, 3, 4, ...$$

Balmer Series
$$1/\lambda = R[1/2^2 - 1/n^2] \quad n = 3, 4, 5, ...$$

Paschen Series
$$1/\lambda = R[1/3^2 - 1/n^2] \quad n = 4, 5, 6, ...$$

In the above equations, R is the **Rydberg's Constant. Its value is R = 1.097 × 10⁷ m⁻¹.**

The Bohr model of the atom is successful in describing the spectra of atomic hydrogen and hydrogen-like ions. One of the basic assumptions of the model is that the electron can exist only in certain orbits such that its angular momentum, mvr, is an integral multiple of $\hbar$ where $\hbar$ is Plank's constant divided by 2π. Assuming circular orbits and a Coulomb force of attraction between electron and proton, Bohr postulated:

1. The electron can orbit around the nucleus in particular (discrete) orbits only. In these orbits, it does not radiate energy.
2. The allowed orbits r_n are those for which the orbital angular momentum is an integral multiple of h/2π.

$$m v r_n = \frac{nh}{2\pi} \quad\quad 9.6$$

where n is an integer (n = 1, 2, 3, ...), h is Planck's constant and m and v the mass and velocity of the electron respectively.

3. When an electron moves from a higher energy level E_i to a lower energy level E_f, photons are emitted.

$$E_i - E_f = h\upsilon$$

The allowed orbits are called **stable energy levels** or **stationary states**. In these orbits, no energy is radiated. To move to a higher orbit, the electron must acquire energy equal to the energy difference of the two orbits. The higher energy levels are called **excited levels.** The minimum energy needed to leave the atom is called **ionization energy.**

The radii for the Bohr orbits are given by

$$r_n = (5.29 \times 10^{-11} m) \frac{n^2}{Z}$$

The radii are quantized.

For hydrogen $r_n = (5.29 \times 10^{-11} m) n^2$

And energy for n = 1 is equal to En = − 13.6 eV.

X-rays are produced when high energy electrons are suddenly decelerated. If the accelerating voltage is V, the electrons acquire kinetic energies KE equal to

$$KE \ = \ \tfrac{1}{2} \, mv^2 \ = \ eV$$

where e is the electronic charge, m the mass and v the speed of the electrons.

The x-ray spectrum has a **continuous spectrum called Bremsstrahlung** (braking radiation) and sharp peaks called **characteristic x-rays.**

The Heisenberg's Uncertainty Principle states that there is a fundamental limitation to measuring accurately the position and momentum of a particle simultaneously. This uncertainty in measurement is given by

$$\Delta x \, \Delta p_x \ \geq \ \frac{h}{2\pi}$$

Lasers are produced by stimulated emission; and the emitted light is amplified. LASER is an acronym of **L**ight **A**mplification by **S**timulated **E**mission of **R**adiation.

When an electron moves from a higher to a lower energy level, radiation is emitted. In a gas, hundreds of electrons may spontaneously move from one energy level to another, producing light photons in different directions. In **Stimulated Emission**, more such electrons are **forced** to jump to lower orbits and a more concerted light is produced. In stimulated emission, the gas is bombarded with photons which trigger electrons into changing their energy levels. In this way, instead of one, two photons are emitted for each electron jump.

Stimulated emission has the following characteristics:

1. The emitted photon and the incoming photon have the same direction.

2. The two photons are coherent. They have the same phase.

3. One photon goes in and two come out for each transition, i.e. amplification occurs.

The laser beam is characterized by:
1. It is monochromatic.
2. It is coherent – all waves are in phase.
3. It does not diverge.
4. It is extremely intense.

Questions

1. How does the energy difference between adjacent energy levels vary with the quantum number n in the Bohr model of the hydrogen atom?
2. Why does the hydrogen spectrum have many lines when the hydrogen atom has only one electron?
3. Which series of the hydrogen spectrum lies in the ultraviolet region?
4. What is the numerical value of the ground state energy level for hydrogen atom?
5. What is the difference between excitation and ionization?
6. How can you distinguish x-rays from gamma rays?
7. X-rays are high energy photons. What is their rest mass?
8. K_α x-rays are produced by which transition of electrons?
9. What do you mean by Bremsstrahlung?
10. What is the ratio of the diameter of the nucleus and atom?
11. How is laser light different from ordinary light?
12. What is coherent light?
13. Discuss the importance of population inversion in the production of lasers?
14. What are metastable states?
15. What kind of image is produced from a hologram?

Problems

1. The Paschen Series is obtained when an electron in the hydrogen atom jumps from a higher orbit to a orbit where quantum number n is
   a) 5 b) 4
   c) 3 d) 2
2. How much energy is emitted by a hydrogen atom, when an electron shifts from the fourth orbit to the third orbit?
3. The radius of the first Bohr orbit is 0.53 A°. The radius of the second Bohr orbit is
   a) 1.06 A° b) 2.12 A°
   c) 0.275 A° c) 0.133 A°
4. What is the radius of the fourth orbit of the hydrogen atom according to Bohr's theory?
5. What is the circumference of the fifth orbit of the hydrogen atom according to Bohr theory?
6. How many de Broglie wavelengths will fit into the sixth orbit of the electron around the hydrogen atom?
7. What is the energy difference for n = 6 to n = 1 in the Lyman Series?
8. What is the energy difference for n = 6 to n = 2 in the Balmer Series?
9. Find the shortest wavelength photon emitted in the Balmer Series (Hint: n = ∞ to n = 2 transition).
10. What is the wavelength of the fourth Balmer line?
11. Find the longest wavelength photon emitted in the Paschen Series.
12. What voltage is required to generate x-rays with a wavelength of 5×10^{-11}m?
13. An x-ray of wavelength 7.1×10^{-11} m is generated. To what voltage difference does this wavelength correspond?
14. An electron falling from the M shell into K shell of the molybdenum atom generates an x-ray of wavelength 6.3×10^{-11}m. What is the energy of this x-ray in Joules and electron-volts?
15. An electron moves i) from n = 3 to n = 1 and ii) from n = 4 to n = 2 level in hydrogen atom. In both transitions, photons are emitted. Which photon has higher frequency?

10 Nuclear Physics

OBJECTIVES

After studying this chapter, the student should be able:
- to understand the terms atomic mass number, atomic charge, unified mass unit.
- to understand isotopes.
- to understand that the mass spectrograph can be used to detect and quantify different masses.
- to understand mass defect and binding energy and their importance in fission and fusion.
- to understand radioactivity, nuclear transmutations and half-life.
- to understand how radiations and matter interact.
- to understand the working of different radiation detectors – Wilson Cloud Chamber, Geiger – Muller Counter and Solid State Detector.
- to understand under what conditions nuclear reactions can occur.
- to understand fission, chain reaction.
- to understand the requirements for nuclear reactors.
- to understand the two main types of nuclear reactors.
- to understand fusion.
- to understand the harmful effects of exposure to radiations.
- to understand the biological effects of radiations and the units to measure radiation.
- to understand uses of radiation.
- to understand basic forces of nature and the particles (hadrons, lepton and quarks) accepted today as the basic building blocks of matter.

10.1 Introduction

The development and progress in nuclear physics started with Becquerel's discovery of radioactivity. This was followed by Rutherford's famous experiment on the scattering of alpha particles by gold foil establishing the concentration of the positive charge in the nucleus. The discovery of the neutron came later in 1932. It was followed by the discovery of artificial radioactivity by the Curies and later nuclear fission by Hahn and Strassman in 1938.

10.2 The Atomic Nucleus

The picture of the atom with a highly positively charged, dense nucleus surrounded by electrons was given by Rutherford in 1911. A few years later, Chadwick discovered the neutron. The atomic model accepted today is of a nucleus in the centre of the atom in which all the mass (approximately 99.9%) of the atom is concentrated surrounded by orbiting electrons (Figure 10.1). The nucleus has protons (positively charged particles) and neutrons (neutral particles) of nearly equal mass. Their numbers are in the same ratio for the lighter elements. As one moves up the periodic table, the number of neutrons progressively exceeds the number of protons. For example, helium He^4 has two protons and two neutrons; phosphorus P^{31} has 15 protons, 16 neutrons; and Uranium U^{235} has 92 protons but 143 neutrons. Neutrons and protons are collectively called **nucleons.**

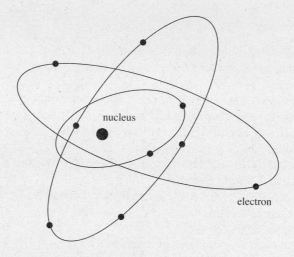

Fig. 10.1: An atom.

The proton is 1836.16 times and the neutron 1838.68 times more massive than the electron. Their masses are measured in **unified mass unit u.** A unified mass unit u is 1/12th the mass of a carbon atom. Carbon is exactly 12 u.

$$\text{One unified mass unit u} \quad = \quad 1.661 \times 10^{-27} \text{ kg} \qquad 10.1$$

Table 10.1 gives the masses of the three elementary particles in unified mass units and in kilograms.

Table 10.1: The masses of the elementary particles

Particle	unified mass units u	kilograms kg
Electron	0.00055	9.11×10^{-31}
Proton	1.007276	1.67×10^{-27}
Neutron	1.008665	1.68×10^{-27}

The atomic mass number gives the total number of protons and neutrons in the atom and is symbolized by A. The atomic number Z gives the number of protons and (A – Z) is the number of neutrons N.

$$\text{Mass number A} \quad = \quad Z + N \qquad 10.2$$

The charge on an electron is equal and opposite of the charge on a proton. The number of electrons and protons in the atom are equal. Thus, the atom is an electrically neutral body. If the charge on one proton is **e,** then the **total nuclear charge** would be **Ze**.

All elements have protons and neutrons in their nucleus with the exception of hydrogen which has only one proton. Nuclei are described by
1) The **atomic number Z**, which is the number of protons in the nucleus.
2) The **neutron number N**, which is the number of neutrons in the nucleus.
3) The **mass number A**, which is the number of nucleons in the nucleus.

Nuclei radii are given by

$$r \quad = \quad r_o A^{1/3} \qquad\qquad 10.3$$

where
$$r_o \quad = \quad 1.2 \times 10^{-15} \text{ m} \qquad\qquad 10.4$$

$$= \quad 1.2 \text{ femtometre (fm)} \qquad\qquad 10.5$$

and A is the atomic mass number.

A **femtometre** is also called a **fermi and is equal to 10^{-15} m.**

Nuclei have tightly packed nucleons and all nuclei have nearly equal densities.

Example: Find the radius and volume of radium-226

$$R \quad = \quad r_o A^{1/3}$$

$$R \quad = \quad (1.2\text{fm}) (226)^{1/3}$$

$$= \quad (1.2 \text{ fm}) (6.091) \text{ fm}$$

$$= \quad 7.31 \text{ fm}$$

$$\text{Volume} \quad = \quad 4/3 \ \pi \ r^3.$$

$$= \quad 4/3 \ \pi \ (7.31 \text{ fm})^3$$

$$= \quad 1636.21 \text{ fm}^3$$

10.3 Isotopes

The symbol $_ZX^A$ represents nuclei, where Z is the charge, A the mass and X the symbol for the element. For example, $_3Li^7$ represents Lithium (Li) having three protons and mass seven, $_0n^1$ is neutron (Z = 0, A = 1) and oxygen $_8O^{16}$ has 8 protons and mass 16.

Elements are known by their atomic number Z. A certain element may have different number of neutrons but the atomic number (the number of protons) remains the same. Nuclei that have the same atomic number Z but different mass numbers A are known as **isotopes. Isotopes of an element have the same atomic number Z but have different mass numbers and neutron numbers.**

For example, carbon has four isotopes $_6C^{11}$, $_6C^{12}$, $_6C^{13}$ and $_6C^{14}$. The more familiar isotope $_6C^{12}$ is nearly 99 % abundant. Hydrogen has three isotopes $_1H^1$ hydrogen, $_1H^2$ deuterium and $_1H^3$ tritium and chlorine has two isotopes $_{17}C^{18}$ and $_{17}C^{20}$.

The chemical properties of isotopes of an element are the same. Other properties are different. For example, the isotope deuterium $_1H^2$ of hydrogen is stable but tritium $_1H^3$ is radioactive and finally changes into an isotope of helium. Heavy water used in reactors is deuterium oxide formed when deuterium atoms combine with oxygen atoms.

Isotopes cannot be separated chemically. They can be physically separated using a mass spectrograph.

10.3.1 Mass Spectrograph

The mass spectrograph has been described in chapter 4. It is found that when charged particles are projected perpendicular to a magnetic field they move in circles given by

$$r \quad = \quad \frac{mv}{qB}$$

If the charges are the same and they are projected with equal velocities **v** into the same magnetic field **B**, then their trajectory radii will be proportional to m

$$r \quad \alpha \quad m$$

This is the principle of the mass spectrograph. Take an element which has two isotopes. Let the ions of the element be accelerated through the same potential difference V. The kinetic energy acquired by the ions is

$$\tfrac{1}{2} mv^2 \quad = \quad q V$$

$$v^2 \quad = \quad (2 q V)/m$$

$$v \quad = \quad \sqrt{\frac{(2qV)}{m}}$$

Since the ions move in circles, the centripetal force is provided by the magnetic force which is given by

$$\mathbf{F} \quad = \quad q \mathbf{v} \times \mathbf{B}$$

$$= \quad q v B \qquad \text{as } \mathbf{v} \text{ perpendicular to } \mathbf{B}$$

Therefore, the centripetal force

$$mv^2/r \quad = \quad q v B$$

$$m \quad = \quad \frac{q B r}{v}$$

Putting in value of v

$$m \quad = \quad \frac{q B r \sqrt{m}}{\sqrt{(2 q V)}}$$

Squaring

$$m^2 \quad = \quad \frac{q^2 B^2 r^2 m}{2 q V}$$

$$m \quad = \quad \left[\frac{q r^2}{2V}\right] B^2 \qquad\qquad 10.6$$

If the value $(qr^2)/2V$ is kept constant, then detector output is proportional to B^2:

$$m \quad \alpha \quad B^2 \qquad\qquad 10.7$$

The field B is adjusted for each ion so that it enters the detector. A graph between B^2 and the detector output will give the relative amounts of each isotope as shown in Figure 10.2.

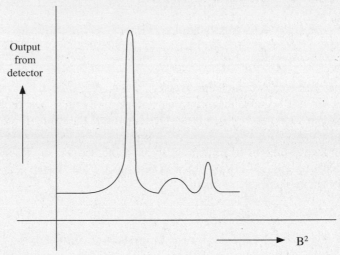

Fig. 10.2: The amount of each isotope of neon (from the right Ne 20, Ne 21 and Ne 22) can be observed from the graph between output and B^2.

Let us consider an element with two isotopes. The atoms of the element are first vapourized and ionized in the ion source. The positively charged ions are accelerated through a potential difference V and projected perpendicular to magnetic field **B**. They move in circular paths (Figure 10.3). Only those ions whose trajectories are of radius r will enter the fixed detector and their number per unit time is measured by the detector. The field is changed so that the second type of ions can now enter the detector and their amount measured. By measuring the relative amounts, the percentage of each in the sample can be calculated.

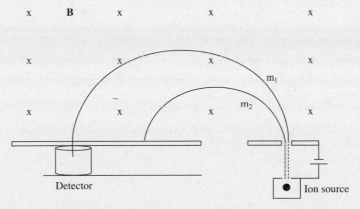

Fig. 10.3: Mass spectrograph separating 2 isotopes of an element.

10.3.2 Nuclear Force

Have you ever thought how so many protons can remain together in the nucleus? How does the nucleus remain stable?

Nuclear stability is because of the existence of the **Nuclear Force.** It is also called **Strong Force.** This force has a short range of about 2×10^{-15} m. It exists between nucleons, that is, between proton–proton, neutron–neutron, and proton–neutron. It is attractive and is stronger than the Coulomb's repulsive force between protons in nuclear dimensions.

$$F_{strong} \quad = \quad \text{highly attractive force} \qquad r \leq 10^{-14} \text{ m.}$$

$$F_{strong} \quad = \quad 0 \qquad\qquad\qquad\qquad\quad r > 10^{-14} \text{ m.}$$

where r is the distance between two nucleons.

As mentioned in the last section, the proportion of neutrons increases progressively for higher elements. This increase in the number of neutrons in turn increases the strong force so that nuclei remain stable (Figure 10.4).

Elements with number of protons more than 83 are unstable because in these the electrostatic force cannot be counter-balanced by the strong force. These nuclei are unstable and disintegrate. They continue to decay till stable nuclei are formed.

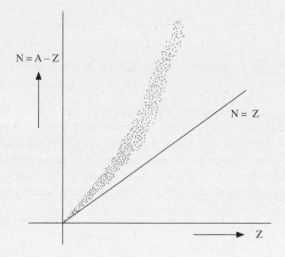

Fig. 10.4: Stable nuclei have number of neutrons N equal or greater than the number of protons Z.

10.4 Binding Energy and Mass Defect

The total mass of the nucleus is less than the sum of the masses of the individual protons and neutrons that constitute the nucleus. This difference in mass is called **mass defect.**

Mathematically it is given by

$$\Delta m \quad = \quad \text{Mass}_{nucleus} - [\, Z \, (M_p) + (A - Z) \, M_n \,] \qquad\qquad 10.8$$

where Δm is the mass defect, $Mass_{nucleus}$ is the mass of the nucleus, Z is the atomic number, M_p mass of each proton, $(A - Z)$ number of neutrons and M_n mass of a neutron.

For the nucleons to remain in the confines of nuclei, energy must be provided to hold them together (Figure 10.5). This energy is called **Binding Energy.**

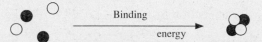

Fig. 10.5: Two protons and two neutrons are bound together to form a helium nucleus.

Using Einstein's mass-energy relationship, the change in mass or the mass defect, Δm, is equal to a change in the energy of the system. The relation is given by

$$\Delta(energy) \quad = \quad (mass\ defect)\ c^2$$
$$\Delta E_o \quad = \quad (\Delta m)\ c^2 \qquad\qquad 10.9$$

where ΔE_o is the change in the rest energy, Δm is the mass defect and c the velocity of light. **The binding energy comes from the mass defect.** This energy must be made available to the nucleus for it to break up into its constituent protons and neutrons.

The mass defect per nucleon is given by

$$\frac{\Delta m}{A} \quad = \quad \frac{Mass_{nucleus} - [Z(M_p) + (A - Z)M_n]}{A} \qquad\qquad 10.10$$

Figure 10.6 is a graph between mass defect per nucleon and atomic number Z. It can be seen from the graph that the maximum values of mass defect per nucleon are from atomic number 25 to atomic number 35. These are the most stable elements.

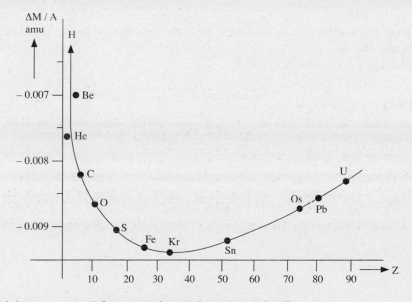

Fig. 10.6: Graph between mass defect per nucleon and atomic number Z.

Example: The helium nucleus has a mass of 6.6447×10^{-27} kg. Find the mass defect and binding energy in joules and eV. Helium nucleus has 2 protons and 2 neutrons.

$$\text{Mass of proton } M_p \quad = \quad 1.6726 \times 10^{-27} \text{ kg}$$

$$\text{Mass of neutron } M_n \quad = \quad 1.6749 \times 10^{-27} \text{ kg}$$

$$\text{Mass}_{nucleus} \quad = \quad 6.6447 \times 10^{-27} \text{ kg}$$

$$\text{Velocity of light c} \quad = \quad 3 \times 10^8 \text{ m/s}$$

The mass defect is

$$\Delta m \quad = \quad (2M_p + 2M_n) - \text{Mass}_{nucleus}$$

$$\Delta m \quad = \quad [2(1.6726 \times 10^{-27}) + 2(1.6749 \times 10^{-27} \text{ kg})] - 6.6447 \times 10^{-27} \text{ kg}$$

$$= \quad 0.0503 \times 10^{-27} \text{ kg}$$

$$\text{Binding energy E} \quad = \quad (\Delta m) c^2$$

$$= \quad (0.0503 \times 10^{-27} \text{ kg}) (3 \times 10^8 \text{ m/s})^2$$

$$= \quad \mathbf{4.53 \times 10^{-12} \text{ J}}$$

We know that

$$1.6 \times 10^{-19} \text{ J} \quad = \quad 1 \text{ eV}$$

$$4.53 \times 10^{-12} \text{ J} = \quad 2.83 \times 10^7 \text{ eV}$$

$$= \quad \mathbf{28.3 \text{ MeV}}$$

The binding energy per nucleon can be calculated for each element by dividing the binding energy by total number of nucleons A.

The binding energy per nucleon is

$$\text{Binding energy/nucleon} \quad = \quad \frac{(\Delta m)c^2}{A} \qquad\qquad 10.10$$

$$= \quad \frac{\{[ZM_p + (A - Z)M_n] - \text{Mass}_{nucleus}\}c^2}{A} \qquad\qquad 10.11$$

When the graph in Figure 10.7 is plotted between the binding energy per nucleon for stable elements and mass number A, it is observed that the binding energy/nucleon rises to 8.8 MeV and then falls to about 7 MeV. With the exception of the lighter nuclei, most nuclei have binding energies of approximately 8 MeV. The maximum occurs for iron ($_{26}Fe^{56}$) which is the most stable element.

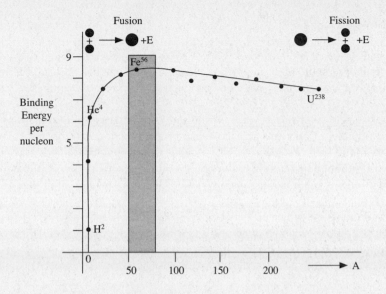

Fig. 10.7: Graph between binding energy per nucleon and mass number A.

The shaded region in the graph marks the most stable nuclei. The curve also explains why fission occurs at one end of the curve and fusion at the other. In both processes the product or products are elements which are more stable elements. Both these processes will be discussed later in the chapter.

The binding energy equivalent for unified mass unit u is equal to

$$\text{Binding energy for u} = (1.661 \times 10^{-27} \text{ kg}) (3 \times 10^8)^2$$

$$= 931.494 \text{ MeV}$$

$$\text{or} \quad 931.5 \text{ MeV} \qquad\qquad 10.12$$

Example: Find the mass defect and binding energy of $_7N^{14}$. The mass of the nucleus is 3.999234 u. The nitrogen atom has 7 protons, 7 neutrons and 7 electrons.

$$\text{Mass defect} = \Delta m$$

$$= [7M_p + 7M_n] - \text{Mass}_{nucleus}$$

$$= (7 \times 1.0072765u + 7 \times 1.00866494u) - 13.9992344u$$

$$= 0.1123564u$$

$$\text{Binding energy} = 0.112356u \times 931.5 \text{ MeV}$$

$$= 104.659 \text{ MeV}$$

10.5 Radioactivity

Radioactivity was discovered by Becquerel in 1896 by accident when he found that uranium salts give radiations that fog photographic films and plates. Madame Curie and her husband Pierre Curie, a couple of years later, isolated successfully two new radioactive elements polonium (Z = 84) and radium (Z = 88).

Radioactive substances emit one or more of the three different types of radiations alpha α, beta β and gamma γ, from nuclei of elements that have an atomic number greater than 83. The elements with Z > 83 have been found to be radioactive. These nuclei are unstable. Radioactivity is a spontaneous process. A nucleus without any warning may emit a helium nucleus (alpha particle), an electron (beta particle) or a high-energy photon (gamma ray). The nucleus changes after the emission of these radiations.

When these radiations are projected between two charged plates, one bends towards the positive plate, the other towards the negative plate and the third passes un-deflected (Figure 10.8). One can conclude that the radiations are negatively charged, positively charged and neutral respectively.

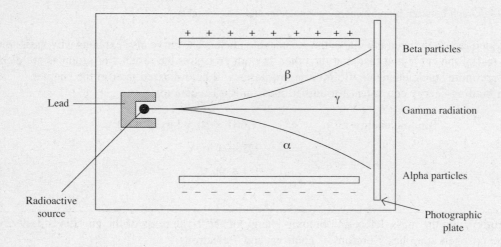

Fig. 10.8: Positive radioactive ions are deflected towards the negative plate, negative towards positive plate and neutral remain un-deflected.

10.5.1 Nuclear Transmutations

Alpha, beta and gamma radiations are emitted from nuclei. With the emission of these radiations, the nucleus changes either into a new nuclide or loses some of its energy and is now less energetic.

In nuclear reactions, the total charge is conserved. The number of nucleons, mass, momentum and energy must be conserved as well. The nuclear reaction is balanced keeping these conservation laws in view.

With the emission of an alpha particle (helium nucleus), the new nucleus has two protons and two neutrons less than the original nucleus.

$$_{Z}X^{A} \quad \rightarrow \quad _{Z-2}Y^{A-4} \quad + \quad _{2}He^{4} \qquad\qquad 10.13$$

The emission of alpha particles usually occurs in the heavier nuclei:

$$_{92}U^{238} \quad \rightarrow \quad _{90}Th^{234} \quad + \quad _{2}He^{4}$$

It can be seen in the above equation that both charge and mass are conserved.

When a beta particle is emitted, the nucleus changes to the next higher element but its atomic mass remains the same. The electron has relative to the nucleus negligible mass which is why the mass number does not change. As its charge is –1, to conserve charge the next higher nucleus in the periodic table must be formed. A general equation for beta emission is

$$_{Z}X^{A} \quad \rightarrow \quad _{Z+1}Y^{A} \quad + \quad _{-1}e^{o} \qquad\qquad 10.14$$

For example

$$_{11}Na^{24} \quad \rightarrow \quad _{12}Mg^{24} \quad + \quad _{-1}e^{o}$$

Electrons do not exist in the nucleus. They are formed at the time of emission of the beta particle.

Gamma rays have no charge or mass. They are high energy photons. They are emitted from nuclei which are in the excited energy state. When these nuclei move to a lower energy level, the energy difference between the two states is emitted as a photon. These photons are called gamma rays.

$$_{Z}X^{A*} \quad \rightarrow \quad _{Z}X^{A} \quad + \quad \gamma \qquad\qquad 10.15$$

The star (*) denotes a nucleus in the excited state.

A positron β^{+} particle may also be emitted by the nucleus.

$$_{Z}X^{A} \quad \rightarrow \quad _{Z-1}Y^{A} \quad + \quad _{+1}e^{o} \qquad\qquad 10.16$$

Both electrons and positrons are produced at the time when a neutron is converted into proton or a proton is converted into a neutron in the nucleus respectively:

$$_{0}n^{1} \quad \rightarrow \quad _{1}H^{1} \quad + \quad _{-1}e^{o} \qquad\qquad 10.17$$

$$_{1}H^{1} \quad \rightarrow \quad _{0}n^{1} \quad + \quad _{+1}e^{o} \qquad\qquad 10.18$$

10.5.2 Radioactive Series

When decay takes place, the original nucleus is called the **parent nucleus** and the nucleus formed after decay is called the **daughter nucleus.** For example, in the decay reaction

$$_{92}U^{238} \quad \rightarrow \quad _{90}Th^{234} \quad + \quad _{2}He^{4} \qquad\qquad 10.19$$

$_{92}U^{238}$ is the parent and $_{90}Th^{234}$ is the daughter nucleus.

Radioactive nuclei keep on disintegrating one after the other till a stable nucleus is produced. This **process of disintegration is called a radioactive series.** There are four such series Thorium, Neptunium, Uranium and Actinium. Of these, the series starting with Neptunium, a trans-uranium ($Z > 92$) element is produced artificially. The radioactive series starting with Thorium is shown in Figure 10.9.

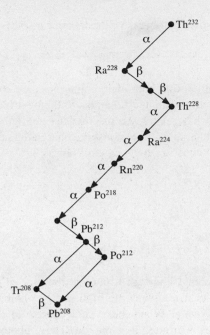

Fig. 10.9: Radioactive decay series starting with Thorium 232.

10.6 Half-life

The decay of nuclei in radioactive substances is a spontaneous process. We do know that so many will disintegrate but we are unable to guess which ones will decay.

As the disintegrations are random, the number of disintegrations per second will be more if more are present. It is proportional to the total number of atoms N present. We can mathematically represent the disintegration rate, also termed activity R as

$$\frac{dN}{dt} \quad \alpha \quad N$$

$$\frac{dN}{dt} \quad = \quad -\lambda N \qquad\qquad 10.20$$

where λ called the **decay constant.** It is the constant of proportionality and dN/dt is the rate of disintegration. The negative sign means that N is decreasing with time. For elements with a larger value of λ, the decay rate is faster.

The rate of disintegration or the activity, that is the number of radiations emitted per second, from sources is measured in becquerels (Bq). **A becquerel is one radiation per second.**

In practice, the activities are high that larger units, **mega-becquerel (10^6 Bq) and giga-becquerel (10^9 Bq)** are employed.

Another larger unit to measure radiations is the **curie (Ci). It is equal to 3.7×10^{10} radiations per second. A curie is 37 GBq.**

As already discussed, radioactive substances do not disintegrate at once. A proportion of the nuclei present keep on decaying. The more nuclei present, the more would decay.

The time in which half the nuclei present decay is called the half-life. Its symbol is $T_{1/2}$.

Let N_0 be the number of nuclei of a radioactive substance present at time t = 0. After time $T_{1/2}$ for that substance ½ (N_0), nuclei will be left. This means that half the nuclei would have disintegrated. After another $T_{1/2}$ time, number of nuclei would become (½) [(½) N_0] and so on. After each half-life, half would be still left. If a graph is plotted between time t and number of nuclei at different times an exponential curve would be obtained (Figure 10.10).

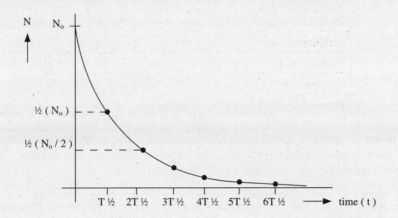

Fig. 10.10: Exponential decay of radioactive nuclei.

For each half-life N_0 is multiplied by 1/2. Thus, after n half-lives the number of nuclei left would be

$$N \quad = \quad (½)^n \, N_0 \qquad\qquad 10.21$$

An equation for N can also be found by integrating

$$\int \; dN/dt \quad = \quad -\lambda N$$

$$N \quad = \quad N_0 \, e^{-\lambda t} \qquad\qquad 10.22$$

The half-life $T_{1/2}$ is the time in which half the nuclei disintegrate and half are left. If at t = 0, the number is N_0. Then after time $T_{1/2}$, the number of nuclei left will be $N_0/2$. Equation 10.22 becomes

$$\frac{N_0}{2} \quad = \quad N_0 \, e^{-\lambda T_{1/2}}$$

$$e^{\lambda T_{1/2}} \quad = \quad 2$$

Taking logarithm to the base e of both sides gives

$$\lambda T_{1/2} \quad = \quad \log 2$$

$$T_{1/2} \quad = \quad \frac{0.693}{\lambda} \qquad\qquad 10.23$$

Different radioactive elements have different half-lives. These range from a few seconds to billions of years. Uranium has a half-life of 4.47 billion years while radon has a half-life of 3.83 days. Table 10.2 gives the half-lives of some important radioactive elements.

Table 10.2: Half-lives of some radioactive elements

Isotope	Half-life	Decay Mode
Polonium $_{84}$ Po ^{214}	1.64×10^{-4} sec	α, γ
Radon $_{86}$ Rn ^{222}	3.83 days	α, γ
Strontium $_{38}$ Sr ^{90}	28.5 years	β^-
Radium $_{88}$ Ra ^{226}	1.6×10^3 years	α, γ
Carbon $_6$ C ^{14}	5.73×10^3 years	β^-
Uranium $_{92}$ U ^{238}	4.47×10^9 years	α, γ

Some isotopes of nuclei, even of Z < 83 exhibit radioactivity. For example, the isotopes iodine-131, carbon-14 and potassium-40 exhibit radioactivity. Others can be made radioactive artificially by bombarding the nuclei with highly energetic particles like neutrons.

Example: The half-life of a radioactive nucleus $_{86}$Ra^{226} is 1.6×10^3 years. Determine the decay constant for radium 226.

$$T_{1/2} = 1.6 \times 10^3 \text{ years}$$
$$= (1.6 \times 10^3 \text{ years}) (3.16 \times 10^7 \text{ s/year})$$
$$= 5 \times 10^{10} \text{ s}$$

Using equation 10.22

$$\lambda = \frac{0.693}{T_{1/2}}$$
$$= \frac{0.693}{5 \times 10^{10} \text{ s}}$$
$$= 1.4 \times 10^{-11} \text{ s}^{-1}$$

Example: Eight gram of radioactive iodine 131 is present. How much iodine would remain after 5 half-lives?

$$N_o = 8 \text{ g}$$

N after 5 half-lives?

From equation 10.20

$$N = (\tfrac{1}{2})^5 N_o$$
$$= (1/32) (8g)$$
$$= 0.25 \text{ g}$$

10.7 Interaction of Radiation with Matter

Radiations (alpha, beta and gamma) from radioactive materials interact with matter in different ways. These include ionization by collision and electrostatic interaction, excitation by losing energy to atoms and molecules which move to higher energy levels and in the case of gamma rays by photoelectric, Compton's effect and pair production effects. The interaction depends on the particle whether it is alpha, beta or gamma and their energies. Ionization depends on these factors as well as the density of the medium and the ionization energy of the atoms.

Alpha particles are positively charged. Alphas being more massive are able to ionize more easily. Because of their size, collision probability increases and they can knock off electrons from the atom or cause electrons to jump to higher orbits. Alphas also produce ions by electrostatic attraction because of their charge of 2e.

Alphas lose energy by ionization and excitation. They can produce 1000s of ion pairs in 1 cm of air. Because of their comparatively high mass, they are not easily deflected by electric and magnetic fields and move in a straight path. They are less penetrating than the other two. They have a short range as they lose energy much faster than the other radiations. A sheet of paper, the skin, a few centimeters of air and aluminium foil of 0.02 mm thickness can stop them. They are particularly dangerous inside the body because of their high ionization probabilities.

Beta particles are electrons so they have negligible mass. Ionization is much less than alpha particles. They produce only 100s of ions. Because of their much smaller size and less ionization probability, their range is much higher. They also lose energy by Bremsstrahlung by producing x-rays. These x-rays can be more penetrating than the beta particles because they have no charge.

The range of beta particles in air is several metres and 1 cm aluminium plate can stop them. The positrons have a very short range. They interact with electrons and annihilate into photons.

Beta particles are deflected by electric and magnetic fields more easily. They also exhibit scattering by outer electrons of atoms.

Beta particles are both more dangerous inside the body as well as outside because unlike alpha particles they can penetrate the skin from outside as well. In a denser material, their range is less.

Gamma rays have no mass and no charge. Their interaction with matter is limited to photoelectric effect, Compton scattering and pair production.

In the **photoelectric effect**, photoelectrons are spontaneously emitted with the absorption of a photon of energy equal or greater than the work function ($h\upsilon_o$) of that atom. Gamma rays of low energies of ≤ 0.5 MeV can produce photoelectrons.

In **Compton scattering**, more energetic gamma rays are required. The radiation is absorbed by an electron in the atom which is excited. It emits a gamma ray of lesser energy.

High energy (≥ 1.02 MeV) gamma rays produce electron-positron pair. Each particle requires 510 keV of energy. This is known as **pair production**. The difference in energy appears as kinetic energy of the electron-positron pair.

Gamma rays like all electromagnetic radiation move in straight lines, are not deflected by electric and magnetic fields and obey the inverse square law. Their intensity in air is inversely proportional to r^2. They are stopped by heavy metals like lead, and concrete. Their intensity in solids reduces exponentially as

$$I \quad = \quad I_o \, e^{-\mu x} \qquad\qquad 10.24$$

Fluorescent materials shine when these rays bombard them. This is due to excitation of the atoms and emission of radiation in the visible spectrum range. This property is used in the detection of the radiations. Photographic films are coated with zinc sulphide, a fluorescent material. Other such substances are barium platinocyanide and sodium iodide.

Neutrons have no charge. They interact with matter through collisions. They can be stopped by protons and hydrogen atoms as they have approximately the same mass. For this reason, neutrons are dangerous as they can interact with water (H_2O) molecules in the human body.

10.8 Radiation Detectors

Different devices and instruments have been designed to detect and quantitatively measure the radiations.

10.8.1 Cloud Chamber

When air is cooled, the water vapour present in the air becomes saturated. It can be cooled even further without any condensation. The air is now **supersaturated** with the vapour. If there are no dust particles, condensation will not take place as the water condenses as droplets on these dust particles.

C.T.R Wilson found that water vapour can condense on gas ions as well. This effect could be used to make the trajectories of charged particles visible.

A glass chamber with a piston and a radioactive source is taken. The air inside is saturated with alcohol vapours which come from a piece of felt dipped in alcohol (Figure 10.11). As the radioactive particles move across the chamber, they ionize the air molecules by removing electrons from the atoms. Along their paths, they leave a trail of positive and negative ions. The piston is pulled down. As the air inside the chamber expands, cooling occurs. The alcohol vapour condenses on the ions. When light is shined on them, the condensed droplets become visible and so the track of the radioactive particle is observed. Photographs can be taken as well.

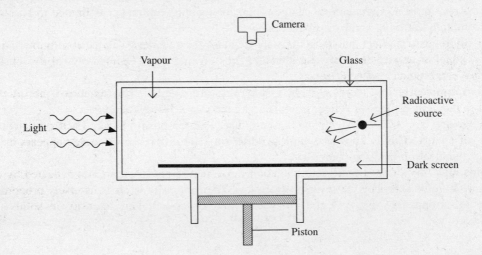

Fig. 10.11: A Cloud Chamber.

The heavier alpha particles are seen to move in straight and thick tracks. They ionize some 10,000 atoms in each centimeter (Figure 10.11a). Beta particles on the other hand being very light and negatively charged are repelled by the electrons of atoms near which they pass. They only make a few hundreds of ion-pairs per centimeter. They move in thin curves (Figure 10.11b).

Gamma rays do not produce tracks themselves. When they ionize by ejecting electrons from air molecules, these electrons behave as beta particles. They move in curved paths like the beta radiations away from the path of the gamma ray (Figure 10.11c).

Fig. 10.11a: Alpha α particle tracks.

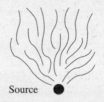

Source

Fig. 10.11b: Beta (β) particle tracks.

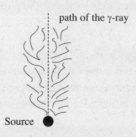

path of the γ-ray

Source

Fig. 10.11c: Gamma (γ) particle tracks.

Cloud chamber studies have proved extremely useful in estimating energies of particles by measuring the length of the tracks. By placing the cloud chamber in a magnetic field, the particle's charge and mass can be determined. A high potential difference across the chamber can be used to clear the chamber of any excess charges.

10.8.2 The Geiger Muller Counter

The Geiger counter uses the ionization of a medium to detect radiations. A cylindrical metal tube (cathode) is filled with argon gas and bromine vapour at low pressure (0.1 atm). A wire electrode (anode) is placed along the axis of cylinder. The potential difference between the wire and the metal tube is kept very high about 1000 V (Figure 10.12). When a charged particle or gamma ray enters the cylinder through a thin mica window on one side of the cylinder, ionization of the gas atoms occurs.

The electrons so produced, accelerate toward the positive central wire. Along their path they ionize other atoms. Thus, an avalanche of electrons is produced. This avalanche of electrons is called **gas amplification.** When the electrons reach the electrodes, a current pulse flows in the circuit. This pulse is amplified and can stimulate an electronic counter or a loud speaker that can make a sound every time a radiation is detected.

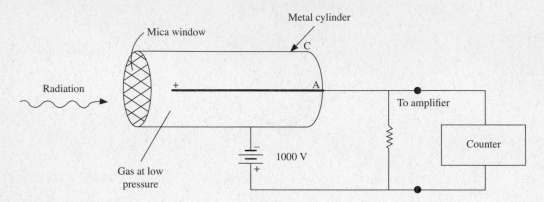

Fig. 10.12: Geiger Counter.

The bromine vapour acts as a **quenching agent.** The vapour absorbs the energy of the positive ions moving towards the cathode (cylinder). Thus, the possibility of any electrons that may be produced when the positive ions strike, the cylinder is eliminated. The Geiger Muller counter will only give a current pulse for one particle that enters the counter.

The time it takes for the slower positive ions to move towards the cathode and be quenched is called the **dead time.** During this time which is approximate 10^{-4} s, the counter is not able to detect another particle.

The Geiger counter cannot be used for particles coming at a rate faster than its dead time.

The counter can also be used to find the range of the radiation in a material. This is done by placing materials of varying thicknesses between the source and the counter (Figure 10.13). The count rate is observed for different thicknesses of the material. From the graph between the thickness and count rate, the range can be easily estimated.

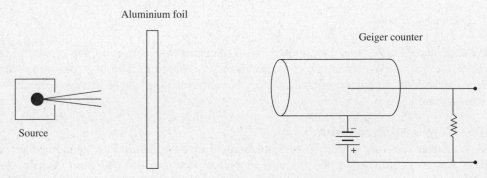

Fig. 10.13: Determination of range of radioactive particles.

10.8.3 Solid State Detector

A semiconductor diode detector is a reverse biased p-n junction as shown in Figure 10.14. Electron-hole pairs are produced when a charged particle passes through the junction. As the electron and hole move away, a current pulse is produced. This pulse activates an electronic counter and so each radiation is detected and counted. The duration of a pulse is 10^{-8} seconds.

In the reverse bias mode, the depletion zone is larger. No current flows through the junction.`

Electron-hole pairs are created as the charged particle goes into the depletion zone. The electrons and holes move in opposite directions towards the junction due to the electric field produced by the reverse biasing potential. The potential is reduced and a small current flows in the circuit. This current is amplified and it then activates a counter connected across the load resistance.

The semiconductor counter can detect weak radiations because very small energies of the order of 3 to 4 eV are needed to produce the electron-hole pairs.

The other advantage is that these detectors are small in size and require no gas filling or high voltages to operate them.

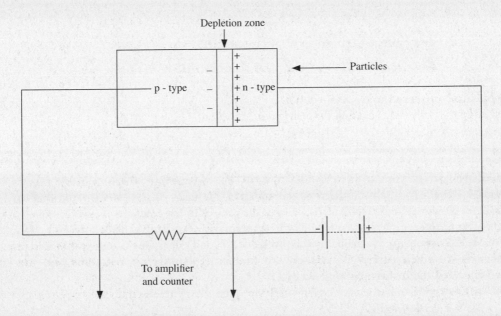

Fig. 10.14: A Semiconductor Detector.

10.9 Nuclear Reactions

Nuclear reactions are the reactions that take place between nuclei and other particles like alpha particle, beta particle, gamma radiation, neutrons and photons. For example, the general equation for the reaction is:

$$X \quad + \quad x \quad \rightarrow \quad Y \quad + \quad y$$

X is the nucleus, x the particle and Y and y the products of the reaction.

When an incident particle or photon changes the target nucleus, the process is called a **nuclear reaction.**

Rutherford in 1919 was able to bombard nitrogen with alpha particles to produce oxygen and a proton. The reaction equation is

$$_2He^4 \quad + \quad _7N^{14} \quad \rightarrow \quad _8O^{17} \quad + \quad _1H^1 \qquad \qquad 10.25$$

A higher element is produced in the reaction. This is a **transmutation reaction**.

In 1932, Cockcroft and Walton bombarded lithium with high speed protons. The two products had a lower mass number:

$$_3Li^7 \quad + \quad _1H^1 \quad \rightarrow \quad _2He^4 \quad + \quad _2He^4 \qquad \qquad 10.26$$

The charge and mass of the reactants and of the products must be equal.

In nuclear reactions, the number of protons and neutrons must be conserved. For example, in the reaction

$$_2He^4 \quad + \quad _4Be^9 \quad \rightarrow \quad _6C^{12} \quad + \quad _0n^1 \quad + \quad \text{energy} \qquad 10.27$$

# of protons LHS = 6 and # of protons RHS = 6

# of neutrons LHS = 7 and # of neutrons RHS = 7

Energy is also conserved in nuclear reactions.

Analyzing the following reaction for energy conservation

$$_1H^2 \quad + \quad _7N^{14} \quad \rightarrow \quad _6C^{12} \quad + \quad _2He^4 \qquad \qquad 10.28$$

Masses of $_1H^2$ and $_7N^{14}$ are 2.014102 u and 14.003074 u respectively. The sum of the masses on the LHS equals 16.017176 u. On the RHS of the equation, the mass of $_6C^{12}$ is 12 u and that of $_2He^4$ is 4.002602 u which totals 16.002602 u. The sum of the masses of the reactants is greater than that of the products. The difference in mass is 0.014574 u. Applying $E = \Delta mc^2$ gives the equivalent energy equal to 13.57 MeV. This energy is carried by the products $_6C^{12}$ and $_2He^4$. This is an **exothermic reaction.**

Reactions in which energy is released are known as exothermic reactions and the energy released is called the disintegration energy Q.

If the incident particle had kinetic energy of 3 MeV, the energy released in the above reaction would be 13.57 MeV + 3 MeV = 18.57 MeV.

In the reaction (equation 10.25)

$$_2He^4 \quad + \quad _7N^{14} \quad \rightarrow \quad _8O^{17} \quad + \quad _1H^1$$

the mass before reaction is 4.002602 u + 14.003074 u = 18.005676 u. After the reaction it is 16.999133 u + 1.007825 u = 18.006958 u. The mass of the reactants is less than that of the products. The difference in mass is 0.001282 u which equals 1.194 MeV. The disintegration energy is –Q. It is an **endothermic reaction**. The reaction will not take place unless energy greater than 1.194 MeV is supplied.

Why should the energy of the incoming alpha $_2He^4$ be more than 1.194 MeV? The energy has to be more than 1.194 MeV to conserve momentum. With only 1.194 MeV energy, the products would have zero kinetic energy and zero momentum.

In nuclear reactions, momentum has to be conserved as well.

A slow neutron (low KE) may be absorbed by a stationary atom producing products that have appreciable kinetic energies. Where did these energies come from? We know that some mass is converted into energy in certain nuclear reactions. The energy appears as the kinetic energy of the products of the reaction.

The emission of an alpha particle transforms the thorium U^{238} into Th^{234}:

$$U^{238} \quad \rightarrow \quad Th^{234} \quad + \quad He^4$$

The products are less massive than U^{238}:

$$
\begin{aligned}
\text{Mass of } U^{238} \quad &= \quad \text{Total masses of product} \\
\text{Mass of } {}_{92}U^{238} \quad &= \quad \text{Mass of } {}_{90}Th^{234} \; + \; \text{Mass of } {}_{2}He^4 \\
238.0508 \text{ u} \quad &= \quad 234.0436 \text{ u} \qquad + \; 4.0026 \text{ u} \\
&= \quad 238.0462 \text{ u}
\end{aligned}
$$

Subtracting RHS from LHS gives the difference in mass equal to 0.0046 u which is equivalent to 4.3 MeV. The mass energy difference appears as the disintegration energy Q.

10.10 Nuclear Fission

Four German scientists Hahn, Meitner, Strassmann and Frisch are responsible for the discovery of atomic fission in 1939. They found that a slow neutron is absorbed by Uranium 235 ($_{92}U^{235}$). The nucleus after absorbing a neutron becomes $_{92}U^{236}$ which is unstable and lasts only 10^{-17} s. Because of its excess energy, it starts vibrating violently and becomes distorted and elongated. A force of repulsion between the two ends finally causes it to break up into two fragments and a couple of neutrons, releasing a tremendous amount of energy. Figure 10.14a and b depict this reaction which is given by

$$_{92}U^{235} \; + \; {}_{0}n^1 \; \rightarrow \; {}_{92}U^{236^*} \; \rightarrow \; X \; + \; Y \; + \; \text{neutrons} \; + \; \text{energy} \qquad 10.29$$

where X and Y are fission fragments

There are a number of such fission reactions producing different fragments X and Y. One of the fission reactions is

$$_{92}U^{235} \; + \; {}_{0}n^1 \; \rightarrow \; {}_{92}U^{236^*} \; \rightarrow \; {}_{56}Ba^{141} \; + \; {}_{36}Kr^{92} \; + \; 3{}_{0}n^1 \; + \; \text{energy} \qquad 10.30$$

Natural uranium is 99.3 % $_{92}U^{238}$ and only 0.7% $_{92}U^{235}$. The probability, that $_{92}U^{238}$ will capture a neutron and break into smaller nuclei is very slim. Therefore, $_{92}U^{238}$ is not good fissionable material. On the other hand, $_{92}U^{235}$ easily captures a **slow** neutron and breaks into two fragments. Slow nuclei are **thermal nuclei with energies equal or less than 0.04 eV**.

Other fission reactions are:

$$_{92}U^{235} \; + \; {}_{0}n^1 \; \rightarrow \; {}_{92}U^{236^*} \; \rightarrow \; {}_{50}Sn^{129} \; + \; {}_{42}Mo^{104} \; + \; 3{}_{0}n^1 \; + \; \text{energy} \qquad 10.31$$

and

$$_{92}U^{235} \; + \; {}_{0}n^1 \; \rightarrow \; {}_{92}U^{236^*} \; \rightarrow \; {}_{54}Xe^{140} \; + \; {}_{38}Sr^{94} \; + \; 2{}_{0}n^1 \; + \; \text{energy} \qquad 10.32$$

The energy generated in the fission reaction is the difference in the binding energies of the heavy nucleus (uranium) and the intermediate nuclei fission products. Let us estimate roughly the energy that would be released during a nuclear fission when an element of A = 240 breaks up into two fragments of A = 120 each.

From the binding energy graph

Binding energy of A=240 is	=	7.6 MeV per nucleon
Binding energy of A=120 is	=	8.5 MeV per nucleon
Difference	=	0.9 MeV per nucleon

For all 240 nucleons, the total energy produced per fission is nearly 200 MeV.

Thus, the energy released per fission is approximately 200 MeV which is 10^8 times greater than that released in a chemical reaction. The mass of the fission products is less than that of the reacting nuclei. This mass loss is equal to the energy produced in the reaction. In some fission reactions, 3 neutrons are produced, in others 2 to 5. The average number of neutrons produced per fission is 2.5 neutrons.

From Figure 10.7, it can be seen that the binding energy per nucleon for the heavy nuclei is less than the binding energy of those of the intermediate masses. These are more tightly bound and have less mass than the heavier nuclei. The difference in mass between the uranium nucleus and the fission products is converted into energy.

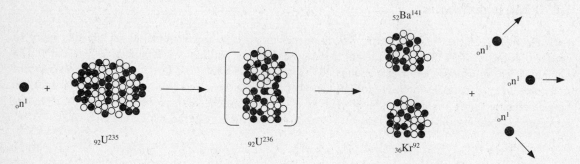

Fig. 10.14a: Fission reaction of Uranium 235 after capturing a slow neutron.

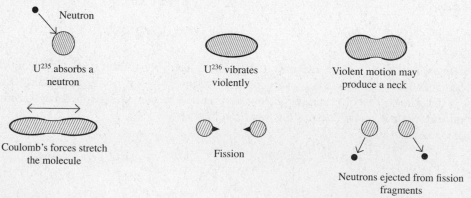

Fig. 10.14b: Fission reaction of Uranium 235 after absorbing a neutron showing different stages of fission.

Example: How much energy is released in the nuclear reaction?

$$_{92}U^{235} + {}_{0}n^1 \rightarrow {}_{56}Ba^{138} + {}_{36}Kr^{95} + 3{}_{0}n^1 + \text{energy}$$

LHS		
Mass of $_{92}U^{235}$	=	235.04390 u
Mass of $_{0}n^1$	=	1.00866 u
	=	236.05256 u
RHS		
Mass of $_{56}Ba^{138}$	=	137.90500 u
Mass of $_{36}Kr^{95}$	=	94.90500 u
Mass of $3\,{}_{0}n^1$	=	3.02599 u
	=	235.83599 u

Difference in mass

$$\Delta m = 236.0526 \text{ u} - 235.8359 \text{ u}$$
$$\approx 0.22 \text{ u}$$

$$\text{Energy released} = 0.22 \text{ u} \times 931.5 \text{ MeV/u}$$
$$= 204.93 \text{ MeV}$$

10.10.1 Chain Reaction

The neutrons produced in the fission reaction can in turn be absorbed by other uranium nuclei which will break into fragments and more neutrons will be produced which will continue to fragment other uranium nuclei. A chain reaction would start, thus, releasing a tremendous amount of energy (Figure 10.15).

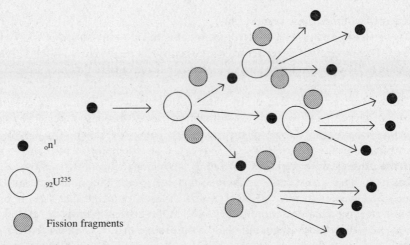

$_{0}n^1$

$_{92}U^{235}$

Fission fragments

Fig. 10.15: A Chain Reaction.

In each fission reaction, three neutrons are produced. These would attack three other uranium nuclei which in turn would produce nine neutrons and so more and more neutrons would fragment and more energy would be produced. This is an **uncontrolled reaction.** An uncontrolled chain reaction can cause an explosion. One kilogram of $_{92}U^{235}$ would release energy equal to 2000 tons of TNT.

If two of these neutrons can be made unavailable and only one neutron keeps the reaction going, then the reaction is a **controlled reaction** (Figure 10.16).

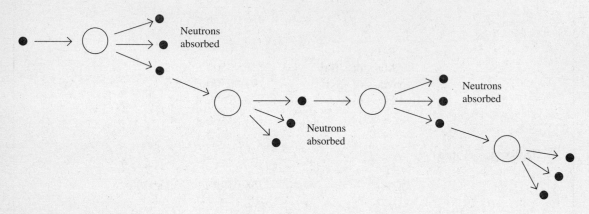

Fig. 10.16: A controlled fission reaction where two nuclei are made unavailable for fission.

10.10.2 Nuclear Reactors

Today there are more than 400 nuclear reactors in 26 countries.

Nuclear reactors are based on controlled chain reactions. A schematic diagram of a nuclear reactor is shown in Figure 10.17. The nuclear reactor is an important application of radioactivity. Energy produced in the reactor is converted into electrical energy for use in industry and homes. At one time, it was thought this energy would solve the world's energy problems but accidents at a number of nuclear facilities made governments wary of these plants. The Three Mile Island Facility in USA and the Chernobyl disaster in Soviet Union are examples of these accidents.

The nuclear reactor has three basic components:
1. **Fuel**: This is the fissile material. The fissile material is made in the shape of rods which are placed close together. They are called the **reactor cores.** A common reactor fuel is $_{92}U^{235}$. As the percentage of $_{92}U^{235}$ is only 0.7% in naturally occurring uranium, its concentration is increased. This is **enriched** fuel and its percentage is 3%.
2. **Neutron Moderator:** Uranium 235 easily absorbs slow neutrons. During the fission reaction, fast neutrons are produced. To slow them down they are passed through water. They collide with the water molecules which absorb some of their energies. They slow down. The slowing process takes 10^{-3} s. They then enter the fuel rods. Water, thus acts as a moderator. **A moderator material slows down neutrons.** Heavy water can also be used as a moderator.
3. **Control Rods:** To have a sustained reaction so that energy is continuously produced, the reaction must proceed at a moderate rate. In other words, it must proceed at a rate of one neutron per fission. At this rate, the chain reaction is said to be in the **critical condition.** Below this value, the reaction is in the **sub-critical stage** and could stop. Above this, it is **supercritical** and may cause an explosion.

Control rods made up of materials that absorb neutrons are used to catch neutrons to slow down the reaction. These rods are inserted between the fuel rods. As the reactor becomes supercritical, the rods are moved further down. To enhance the reaction rate, they are moved out. The rods are made of carbon, boron or cadmium that absorbs neutrons.

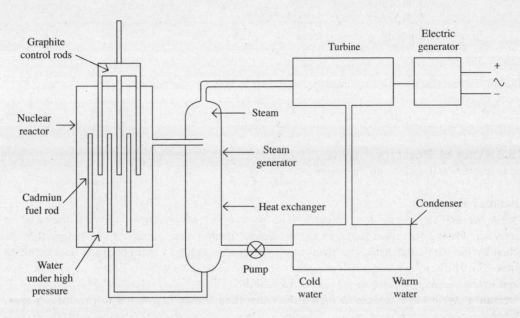

Fig. 10.17: Nuclear Reactor.

In the pressured water reactor (PWR), the reactor vessel has water at high pressure (about 150 atm) as a heat transfer medium. The pressure is kept high so that the water does not boil. The tremendous amount of energy produced in the reactor heats (temperatures about 600 K) up the water which goes to a heat exchanger and steam generator. Here, steam at very high temperatures and pressure is produced. This steam operates turbines that drive electrical generators. The thermal energy is so converted into electrical energy. The used steam then passes into a condenser where it is cooled and condenses into water. The water is pumped back into the steam generator. The cycle continues.

If reaction rate becomes high, the heat produced may melt the reactor core. Exposure to radiation may occur as well. Nuclear reactors must be constructed in areas away from settlements. Disposal of nuclear waste material is a major problem. Today, radioactive material is placed in sealed, thick walled containers which are then stored in underground mines. Transportation of the radioactive fuel material as well as radioactive waste is another threat to people and the environment.

Example: $_{92}U^{235}$ can absorb a neutron and produce fission fragments Xe^{140} and Sr^{94}. How many neutrons are produced?

In a chemical reaction, the two sides of the equation must be balanced.

$$_0n^1 \quad + \quad _{92}U^{235} \quad \rightarrow \quad _{54}Xe^{140} \quad + \quad _{38}Sr^{94}$$

L.H.S	A	=	236
R.H.S	A	=	234

$$\text{Difference in atomic mass} \quad = \quad 236 - 234$$
$$= \quad 2$$

In the above reaction, 2 neutrons must be produced to balance the equation

$$_0n^1 \quad + \quad _{92}U^{235} \quad \rightarrow \quad _{54}Xe^{140} \quad + \quad _{38}Sr^{94} \quad + \quad 2_0n^1$$

10.10.3 Types of Reactors

Nuclear reactors fall under two categories:

i) Thermal Reactors.

The term thermal applies to low energies. The pressurized water reactor (PWR) discussed in the previous section is a thermal reactor. In these reactors, neutrons are slowed down before they can be absorbed by the fissile material. The usual fissile material is natural uranium which is enriched to 3% uranium-235. In these reactors, high pressures are applied to the water or heavy water that transfers the heat to the steam generator to prevent it from boiling.

Thermal reactors are devices in which the controlled fission is utilized to produce energy.

Thermal reactors are more common. The first reactor to produce electricity was built in Idaho, USA in 1951.

ii) Fast Reactors

As the name suggests, fast reactors employ fast neutrons. They are also known as **breeder reactors.** In these reactors, a moderator to slow down neutrons is not needed.

In fast reactors, fissionable material is produced at a greater rate than the rate at which fuel is consumed.

These reactors use plutonium 239, an artificially produced, trans-uranium element as fuel.

When uranium-238 captures a neutron, uranium-239 and a gamma is produced. The uranium-239 isotope is unstable ($T_{1/2}$ = 24 min) and beta-decays to form neptunium-239 which in turn disintegrates ($T_{1/2}$ = 2.3 days) to form plutonium-239 and beta particle:

$$_0n^1 \quad + \quad _{92}U^{238} \quad \rightarrow \quad _{92}U^{239} \quad + \quad \gamma \qquad \qquad 10.33$$
$$\downarrow$$
$$_{93}Np^{239} \quad + \quad _{-1}e^0$$
$$\downarrow$$
$$_{94}Pu^{239} \quad + \quad _{-1}e^0$$

The plutonium-239 is chemically different from uranium-238 and so can be easily separated from it.

The reactor vessel has a mixture of uranium dioxide and plutonium surrounded by uranium-238 which reacts with the fast neutrons and is converted to plutonium-239.

As uranium-238 is 140 times more abundant than uranium-235, breeder reactors are more viable. In nuclear weapons, plutonium-239 is used. The breeder reactors add to complications over the control of nuclear weapons.

10.11 Nuclear Fusion

When two light nuclei combine to form a heavier nucleus, the process is called nuclear fusion.

The very light nuclei have smaller binding energies. As in the fission reaction when these combine to form larger nuclei, energy is released. Four hydrogen nuclei fuse to form a helium nucleus. This nucleus has a mass less than the four hydrogen nuclei forming it. The difference in mass accompanies release of energy according to Einstein's equation

$$\text{Energy} \quad = \quad (\Delta m)c^2$$

In the section on fission it was found that the energy released per nucleon was about 1 MeV. In fusion where two nuclei of very low mass and relatively small binding energies fuse together the energy released is approximately 3.4 MeV. If reactors could be designed to continuously release energy by fusion, this process would solve the world's energy crises with limitless fuel and pollution free energy production! A reactor using fusion has not been constructed as yet due to the high cost of starting such a reaction, controlling and maintaining a continuous fusion reaction.

Fusion can occur by different reactions. When two deuterons combine to form a helium nucleus 24 MeV of energy is released:

$$_1H^2 \quad + \quad _1H^2 \quad \rightarrow \quad _2H^4 \quad + \quad 24 \text{ MeV} \qquad\qquad 10.34$$

Another reaction is to combine two deuterons to form tritium and a proton or to form isotope of helium and a neutron:

$$_1H^2 \quad + \quad _1H^2 \quad \rightarrow \quad _1H^3 \quad + \quad _1H^1 \quad + \quad 4\text{MeV} \qquad\qquad 10.35a$$

$$_1H^2 \quad + \quad _1H^2 \quad \rightarrow \quad _2H^3 \quad + \quad _0n^1 \quad + \quad 3.3\text{MeV} \qquad\qquad 10.35b$$

The two reactions have equal probabilities. A deuterium-tritium mixture can release 17.6 MeV of energy:

$$_1H^3 \quad + \quad _1H^2 \quad \rightarrow \quad _2H^4 \quad + \quad _0n^1 \quad + \quad 17.6 \text{ MeV} \qquad\qquad 10.36$$

It is difficult to start and maintain a fusion reaction. Bringing two similarly charged nuclei together against the electrostatic repulsive force requires very high kinetic energies of the two nuclei. They have to come to distances so close that the nuclear strong forces can then pull them and fusion can take place. The nuclei need to have these high energies so that they can overcome the electrostatic repulsion between two positively charged bodies.

For two deuterium nuclei to fuse together each must have kinetic energy equal to 0.25 MeV. As kinetic energy is related to temperature, this energy corresponds to a temperature of 2×10^9 K. Temperatures that high are impossible to achieve except for a short time in fission. It is for this reason, an atomic bomb is needed to detonate a nuclear bomb.

10.11.1 Nuclear Reaction in the Sun

The sun produces energy equal to 4×10^{26} J/s. It provides to the earth 1.4 kW of radiant energy per m^2 area. The sun is able to provide us energy continuously because of the fusion reaction in the sun. Hydrogen combustion in the sun has been going on for 5×10^9 years.

For fusion to occur in stars, the temperature in the star must be 10^7 K and the density of nuclei must be very high so that collisions can take place.

The following reaction is accepted as taking place in the sun. It is the fusing of four protons to form a helium nucleus and is carried out in three steps. It releases 25 MeV of energy. The energy is approximately 6 MeV per nucleon which is comparatively much higher than that released in fission of 1 MeV per nucleon. The reaction is called the **proton-proton cycle** and is as follows:

$$_1H^1 + {}_1H^1 \longrightarrow {}_1H^2 + {}_{+1}e^\circ$$

$$_1H^1 + {}_1H^2 \longrightarrow {}_2He^3$$

$$_1H^1 + {}_2He^3 \longrightarrow {}_2He^4 + {}_{+1}e^\circ \qquad\qquad 10.37$$

These reactions are called thermonuclear fusion reactions.

10.12 Exposure to Radioactivity

The harmful effects of radiations were not known in the early years after the discovery of radioactivity by Becquerel at the end of the 19th century. Madame Curie suffered from anaemia, had a miscarriage and died of leukaemia. She was exposed to radiations in her laboratory.

Many women, who painted radium on the dials of watches to make them glow in the dark suffered from anaemia and died of bone cancer. They were exposed to alpha radiation because they licked radium from the painting brushes when trying to make the brushes into points.

How radioactivity affects people depends on:
1) the type of radiation
2) the frequency of exposure
3) the amount of exposure and
4) whether the radiation source is inside or outside the body.

Alpha radiation is the most dangerous from inside the body. Radon gas, if breathed into the body can cause irreparable harm as it is an alpha emitter.

Cells that are dividing are more vulnerable to radiation. Especially sensitive cells are those of the bone marrow, reproductive organs, spleen and embryo.

10.12.1 Background Radiation

We cannot avoid radiations completely. We eat, breathe and drink minute amounts of radiation all the time. Some radioactive substances with long half-lives accumulate in the body. Radioactive strontium-90 is taken into the body via food. It dislodges calcium from the bones and takes its place. It is a beta emitter. It has a half-life of 28 years. It can cause leukemia and other cancers. Cesium-137 accumulates in all tissues. It too has a long half-life of 30 years.

The unavoidable radiation to which we are exposed has a number of sources:
1) Cosmic rays from space.

2) Radioactive material in rocks and soil. The building materials like bricks and concrete may contain radioactive material like $_{92}U^{238}$, $_{90}Th^{234}$ and $_{19}K^{40}$.

3) Radon gas can enter rooms from the walls and ground. It is imperative that the rooms are kept properly ventilated to reduce exposure from radioactive gases. In food, air and water radioactivity mostly comes from $_{92}U^{238}$, $_{90}Th^{234}$, $_{19}K^{40}$, $_{86}Rn^{222}$, $_{6}C^{14}$ and $_{1}H^{3}$.

10.12.2 Radiation from Human Activities

Exposure to radiation is also from human-related activities like medical and dental x-rays, living in brick or stone structures, smoking cigarettes, nuclear weapons fallout, occupational exposure, living close to a nuclear power plant, smoke detectors, television or computer screens, industrial and hospital wastes, etc.

Exposure to the small background radiation over a long time may not be harmful. The body has an excellent defense system and can repair itself. But the same exposure in a short time may result in mutations, genetic defects and cancer. It has been determined that lung cancer hardly occurs in non-smokers. The tobacco fertilizer has small amount of radium, which deposits on the tobacco leaves. When cigarette is smoked, this enters the lungs and disintegrates into polonium-210 which is an alpha emitter.

10.13 Biological Effects of Radiation

Ionization is the main reason of concern in biological systems. It produces radicals that are capable of altering cell structure leading to malfunction, interfere with the normal working of the cell or in severe cases, death of a cell. These radicals can interfere in the cell's chemical reactions, break bonds in bio-molecules and cause changes in DNA leading to cancer.

Ultraviolet radiations, x-rays, gamma rays and alpha and beta particles carry enough energy to ionize atoms and molecules. To ionize energies between 1 eV to 100 eV is required. Fast alpha radiations may carry 1 MeV of energy and could ionize thousands of molecules. They are extremely hazardous to the body.

For biological purposes, radiation dose or absorbed dose is the amount of energy absorbed by the absorbing material. Thus, the absorbed dose measures the actual radiation dose absorbed per unit mass of the tissue. The tissue could be a person's chest or leg.

$$\textbf{Absorbed Dose} = \frac{\text{Energy absorbed}}{\text{Mass of absorbing material}}$$

$$\textbf{D} = \frac{E}{m} \qquad\qquad 10.38$$

The unit of absorbed dose is the gray (Gy). It is one joule of energy absorbed per kilogram of the absorbing material

$$1 \text{ Gy} = 1 \text{ J/kg} \qquad\qquad 10.39$$

The **rad,** which is an acronym for **r**adiation **a**bsorbed **d**ose, is a smaller unit. It is equal to 10^{-2} J energy absorbed by one kilogram and is related to the gray by

$$\textbf{1 rad} = \textbf{0.01 gray} \qquad\qquad 10.40$$

A high dose of about 3 Gy may be a lethal dose to 50% of the population. The dose we are exposed to normally in a year is 2 mGy.

Some radiations are more dangerous and cause more damage than others. Neutrons may produce eye cataracts at 1 rad while x-rays are hardly likely to.

The **relative biological effectiveness (RBE)** also termed **quality factor (QF)** is used to compare the biological effect of a certain radiation to that of 200 keV x-rays (which is assigned RBE = 1) that produce the same effect.

For x-rays and beta particles, the RBE factor is 1, for slow neutrons it is 5 and for alpha particles it ranges from 10 – 20. Alphas cause 10 times more damage than x-rays.

To differentiate between the effects of different types of radiations and the biological damage they cause another unit is defined. The biologically equivalent dose or **dose equivalent D_e** is the product of the absorbed dose and relative biological effectiveness:

$$\text{Biologically equivalent dose} \quad = \quad \text{absorbed dose} \times \text{RBE}$$

$$D_e \quad = \quad (D)\,(RBE) \qquad\qquad 10.41$$

The SI unit for dose equivalent is the **sievert (Sv)**. The **rem** is another unit. It is related to the Sv by

$$1\ Sv \quad = \quad 100\ rem \qquad\qquad 10.42$$

A single high dose of radiation can cause radiation sickness. The symptoms are nausea, vomiting, diarrhoea, hair loss. If the dose is large it may be fatal. At 50 rem there are hardly any effects. Long term exposure may result in cancer and genetic mutations. A dose of 400 – 500 rem can be fatal most of the time.

Example: A 70 kg person is exposed to whole body dose of 30 rad. How many joules of energy are given to the person?

$$1\ \text{rad} \quad = \quad 10^{-2}\ \text{J/kg}$$

$$30\ \text{rad} \quad = \quad 30 \times 10^{-2}\ \text{J/kg}$$

$$\text{Energy per kg} \quad = \quad 30 \times 10^{-2}\ \text{J}$$

$$\text{In 70 kg energy is} \quad = \quad 30 \times 10^{-2} \times 70$$

$$= \quad \frac{2100}{100}$$

$$= \quad 21\ \text{Joules}$$

10.14 Uses of Radioactivity

Radiations from radioactive nuclei are extremely dangerous. But these radiations can be used for different purposes which include radiation therapies, radiocarbon dating, smoke detectors, radiography and the use of radio-isotopes as tracers.

10.14.1 Radiocarbon Dating

Radiocarbon dating is related to radioactive carbon-14 (C^{14}). Carbon-14 is produced when cosmic rays enter the upper atmosphere. They react with the air molecules and neutrons among other particles are formed. Nitrogen-14 is converted into carbon-14. The reaction is:

$$_7N^{14} \ + \ _0n^1 \ \rightarrow \ _6C^{14} \ + \ _1H^1 \qquad\qquad 10.43$$

The percentage of C^{14} is very small. The ratio of C^{12} and C^{14} remains a constant in living things. When a living organism or plant dies, the ratio of C^{14} to C^{12} decreases as C^{14} starts disintegrating. It has a half-life of 5730 years. By measuring the disintegration rate the age of wood samples, charcoal, fossils and skeletons has been determined.

10.14.2 Smoke Detectors

Smoke detectors are mostly ionization types which use weak radioactive source. The radiations from the radioactive source ionize the air in the ionization chamber. Ionization takes place continuously and a small current is maintained in the detector. When smoke particles enter the chamber, they slow down the ions. The current through the detector decreases. This decrease in current activates the alarm (Figure 10.18).

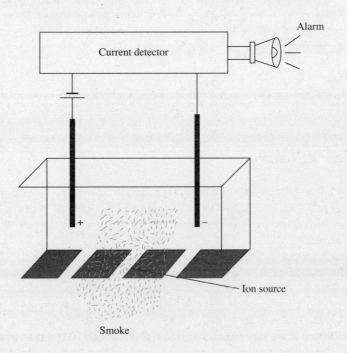

Fig. 10.18: Smoke Detector.

10.14.3 Radioactive Tracers

Radioactive isotopes are used as tracers to detect clogged arteries, sites of internal haemorrhage inside the body, bone problems, growth in plants, etc.

Radio isotope sodium is injected into the blood stream. As it moves down the blood vessel, detectors check its progress. If at a certain point the blood vessel is clogged, radiations would stop or decrease and so the blockage can be pinpointed.

The thyroid gland distributes iodine to different parts of the body. The condition and working of the thyroid gland can be monitored using radioactive iodine-131 that is artificially produced and given to the patient. If the gland absorbs more or less than an optimum amount of iodine, it is malfunctioning. The radiations from iodine-131 can also be used to kill cancer cells in the thyroid gland.

Gallium-67 goes to certain types of tumour and is so useful in their detection.

Gamma radiations are used in radiography. Accurate location of tumours in the brain or body can be determined by taking different images at various angles, feeding the data into computers and forming three dimensional images.

Growing and dividing cells are particularly susceptible to radiations. Thus, radiations can be used to kill cancer. Other would be less affected. For doing this a beam of gamma particles from Co-60 are focused on the tumour. Careful regulation of the dosage and monitoring must be carried out so as not to destroy more than necessary normal cells as well affect personnel operating the radiation facility.

Radioactive material is used in industry. To check leakage in pipes, radioactive gas may be passed through them. Detectors placed above the pipes are used to determine where the leak is. Car engine parts are made of radioactive material. The oil from the engine can be checked for radioactivity. The wear and tear of the engine can be determined by the amount of radioactivity measured.

Plant growth for different nutrients can be checked by using radioactive nitrogen, potassium, sodium, etc. Using these tracers scientists have been able to determine which element is required where in the plant and at what time during its growth.

10.14.4 Sterilization

Surgical instruments can be sterilized using radiations. Gamma radiations from cobalt-60 are used to kill salmonella bacteria in chicken, bacteria on fruits and vegetables and insect eggs in wheat.

10.15 Basic Forces in Nature

Particles in nature interact through four fundamental forces:
1. **The strong force** binds quarks together to form neutrons and protons. It is also the force that binds nucleons to form nuclei. It is the strongest force and its range is 10^{-15} m. At distances greater than 10^{-15} m it is negligible.
2. **The electromagnetic force** is hundred times weaker than the strong force. It is responsible for binding atoms and molecules. It is an inverse square force. Its magnitude decreases with distance. It is a long-range force.
3. **The weak force** is also a nuclear force. It is a short range force that produces instability in certain nuclei. It is responsible for beta decay and is $1/10^6$ times the strong force.
4. **The gravitational force** is a long-range force and is the weakest of the four forces. It is a force that holds the planets, stars, solar system and galaxies together. It is 10^{-43} of the strong force.

These forces are mediated through field particles—**photons** for electromagnetic interaction, **gluons** for strong force, z bosons for the weak force and graviton for the gravitational force. With the exception of the graviton, all other field particles have been detected (Table 10.3).

Table 10.3: Particle Interactions

Force	Strength	Range	Particle
Strong	1	1 fm	gluon
Electromagnetic	10^{-2}	$\alpha\ 1/3r^2$	photon
Weak	10^{-6}	10^{-3} fm	z boson
Gravitational	10^{-43}	$\alpha\ 1/r^2$	graviton

Abdus Salam, Steven Weinberg, Sheldon Glasgow in 1979 won the Nobel prize for formulating the **electroweak theory** that unified the electromagnetic and weak forces. The two forces are manifestations of the same one force—electroweak force. It is believed that all these four forces evolved from a Unified Force after the Big Bang 15–20 billion years ago (Figure 10.19).

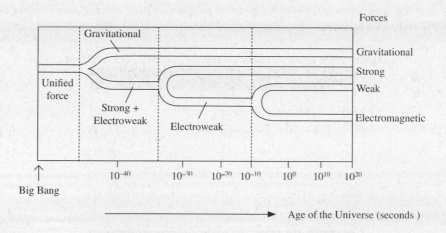

Fig. 10.19: The four fundamental forces and their origin.

10.16 Building Blocks of Matter

Particles are categorized into three groups: photons, leptons and hadrons. This grouping is according to the type of force by which they interact with each other.

1. Photons

There is only one particle (photon) in this group. Photons interact with charged particles and the interacting force is the electromagnetic force.

2. Leptons

Leptons are particles like the electrons that interact via the weak nuclear force. They can also exert gravitation and electromagnetic forces. There are six leptons discovered. These include muon, tau and neutrino.

3. Hadrons

Hadrons interact through the strong nuclear forces as well as weak nuclear forces. They can also interact by electromagnetic forces but only in nuclear distances. Included in hadrons are the proton, neutron, and pions. Hadrons form **mesons** and **baryons**.

a) Mesons decay into electrons, positrons, neutrinos and photons. There are different types of mesons: pion, K meson. The pion (π°) is the lightest of the mesons.

b) Baryons have masses equal or greater than the proton.

10.17 Quarks

A more recent theory suggests that hadrons are complex particles and have a substructure. Hadrons are made of two or three fundamental particles called quarks and anti-quarks. In 1963, Gell-Mann and George Zweig proposed that mesons and baryons were made up of two or three quarks. **Quarks have fractional charges**.

Mesons are made up of quarks and anti-quarks while baryons are made up of three quarks. Originally there were three quarks with interesting names—up (u), down (d) and sideways (also called strange s). All hadrons contain u and d. A proton consists of three quarks—two up, one down while a neutron is made of two down quarks and one up quark (Figure 10.20).

Table 10.3: Different quarks and anti-quarks with charges.

Quark	Charge
u	+ 2/3 e
d	− 1/3 e
c	+ 2/3 e
s	− 1/3 e
t	+ 2/3 e
b	− 1/3e

Anti-quark	Charge
u	− 2/3 e
d	+ 1/3 e
c	− 2/3 e
s	+ 1/3 e
t	− 2/3 e
b	+ 1/3e

New discoveries and fancy names herald more quarks like truth t, charm c and beauty b (Table 10.3). The physics of these particles is still in its infancy stage.

Proton   Neutron   ○ Down quark   ● Up quark

Fig. 10.20: A proton and neutron showing constituent quarks.

Summary

Nuclei are represented symbolically as $_Z X^A$, where X represents the chemical symbol for the element. The quantity A is the **mass number,** which equals the total number of nucleons (neutrons plus protons) in the nucleus. The quantity Z is the **atomic number,** which equals the number of protons in the nucleus.

Nuclei that contain the same number of protons but different numbers of neutrons are called **isotopes.** In other words, isotopes have the same Z value but different A values.

Most nuclei are approximately spherical, with an average radius given by

$$r = r_o A^{1/3}$$

where A is the mass number and r_o is a constant equal to 1.2×10^{-15}m.

The total mass of a nucleus is always less than the sum of the masses of its individual nucleons. This mass difference, Δm, multiplied by c^2 gives the **binding energy** of the nucleus.

The spontaneous emission of radiation by certain nuclei is called **radioactivity.** There are three processes by which a radioactive substance can decay: alpha (α) decay, in which the emitted particles are $_2 He^4$ nuclei; beta (β) decay, in which the emitted particles are electrons or positrons; and gamma (γ) decay, in which the emitted particles are high-energy photons. With the emission of these radiations, the nucleus changes either into a new nuclide or loses some of its energy and is now less energetic.

The **decay rate,** or **activity**, R, of a sample is given by

$$R = \frac{\Delta N}{\Delta t} = -\lambda N$$

where N is the number of radioactive nuclei at some instant and λ is a constant for a given substance called the **decay constant.**

Nuclei in radioactive substance decay in such a way that the number of nuclei present varies with time according to the expression

$$N = N_o e^{-\lambda t}$$

where N is the number of radioactive nuclei present at time t, N_o is the number at time t = 0 and e = 2.718...

The **half-life,** $T_{1/2}$, of a radioactive substance is the time required for half of a given number of radioactive nuclei to decay. The half-life is related to the decay constant as

$$T_{1/2} = \frac{0.693}{\lambda}$$

In nuclear reactions, the total charge mass and energy are conserved. The nuclear reaction is balanced keeping these conservation laws in view.

With the emission of an alpha particle (helium nucleus), the new nucleus has two protons and two neutrons less than the original nucleus.

$$_Z X^A \rightarrow _{Z-2} Y^{A-4} + _2 He^4$$

The emission of alpha particles usually occurs in the heavier nuclei:

$$_{92} U^{238} \rightarrow _{90} Th^{234} + _2 He^4$$

When a beta particle is emitted, the nucleus changes to the next higher element but its atomic mass remains the same.

$$_Z X^A \rightarrow _{Z+1} Y^A + _{-1} e$$

For example

$$_{11} Na^{24} \rightarrow _{12} Mg^{24} + _{-1} e$$

Electrons do not exist in the nucleus. They are formed at the time of emission of the beta particle.

Gamma rays have no charge or mass. Gamma ray emission is given by

$$_Z X^{A*} \rightarrow _Z X^A + \gamma$$

A positron β^+ particle may also be emitted by the nucleus.

$$_Z X^A \rightarrow _{Z-1} Y^A + _{+1} e^o$$

Both electrons and positrons are produced at the time when a neutron is converted into proton or a proton is converted into a neutron in the nucleus respectively:

$$_0n^1 \rightarrow {}_1H^1 + {}_{-1}e^0$$

$$_1H^1 \rightarrow {}_0n^1 + {}_{+1}e^0$$

When decay takes place, the original nucleus is called the **parent nucleus** and the nucleus formed after decay is called the **daughter nucleus.** For example, in the decay reaction

$$_{92}U^{238} \rightarrow {}_{90}Th^{234} + {}_2He^4$$

$_{92}U^{238}$ is the parent and $_{90}Th^{234}$ is the daughter nucleus.

Radioactive nuclei keep on disintegrating one after the other till a stable nucleus is produced. This **process of disintegration is called a radioactive series.** There are four such series, Thorium, Neptunium, Uranium and Actinium.

Nuclear reaction general equation is:

$$X + x \rightarrow Y + y$$

X is the nucleus, x the particle and Y and y the products of the reaction.

Nuclear reactions in which energy is released are said to be **exothermic reactions** and are characterized by positive Q values. Reactions with negative Q values, called **endothermic reactions,** cannot occur unless the incoming particle has atleast enough kinetic energy to overcome the energy deficit.

Nuclear reactors fall under two categories:
i) Thermal Reactors
ii) Fast Reactors

The **absorbed dose** measures the actual radiation dose absorbed per unit mass of the tissue. The tissue could be a person's chest or leg.

$$\text{Absorbed Dose} = \frac{\text{Energy absorbed}}{\text{Mass of absorbing material}}$$

$$\mathbf{D} = \frac{E}{m}$$

Particles in nature interact through four fundamental forces:
   i)   **The Strong Force**
   ii) **The Electromagnetic Force**
   iii) **The Weak Force**
   iv) **The Gravitational Force**

Particles are categorized into three groups: **photons, leptons and hadrons**.

Hadrons are made of two or three fundamental particles called quarks and anti-quarks. **Quarks have fractional charges**. Isolated quarks are not observed.

Protons and neutrons are not fundamental particles; they are made up of quarks.

Questions

1. What is the neutron/proton ratio for californium $_{98}Cf^{249}$?
2. How many protons and neutrons are in $_{19}K^{40}$?
3. What is meant by subatomic particle?
4. What is meant by annihilation?
5. What are essential features of a Geiger – Muller counter tube?
6. What are two major problems in constructing a fusion reactor?
7. How does the sun generate energy?
8. What are transuranic elements?
9. Why are high energy particles needed to bombard nuclei?
10. How do we know that nuclear forces exist?
11. Define radioactivity. Which elements are naturally radioactive?
12. α particles are less dangerous than β particles outside the body. Why?
13. Can radiocarbon dating be used to determine the age of a gold statue?
14. What is the purpose of a nuclear moderator?
15. Which nuclides have the smallest binding energy per nucleon?
16. Which nuclides have the largest binding energy per nucleon?

17. How does binding energy per nucleon affect the stability of the nucleus?
18. Compare the stability of $_6C^{13}$ and $_7N^{12}$.
19. What do isotopes of an element have in common? How do they differ?
20. How does mass of electron compare to mass of a hydrogen atom? What does this signify?
21. What element has the highest MeV/nucleon?
22. What factors make fusion reactions difficult to start?
23. Name the particles in the lepton family.
24. What is the difference between baryons and mesons?

Problems

1. Which particle can easily penetrate the nucleus of an atom: proton, electron, neutron, deuteron, particle? Give reasons for your choice.
2. Complete the equation: $\lambda\, T_{1/2} = \ln \underline{\qquad} = \underline{\qquad}$
3. Assume in the fission of Uranium 235 when it is bombard with a neutron, two stable nuclides with masses 94.9360 u and 138.9500 u are formed along with 2 neutrons. If the mass of U^{235} is 235.1180 u, what mass is converted into energy? What is the binding energy?
4. The half-life of a certain radioactive nuclide is 1 min. What fraction of the original sample of this isotope is left after 10 min?
5. What is the mass defect for tritium nucleus $_1H^3$ if its mass is 3.01605 u?
6. What is the binding energy and the binding energy per nucleon of tritium?
7. Calculate the mass defect, binding energy and binding energy per nucleon of helium nucleus $_2He^4$ if the isotopic mass for $_2He^4$ is 4.00260 u. Use mass of proton = 1.007825 u and mass of neutron = 1.008665 u.
8. Compute the mass deflect and binding energy per nucleon of $_6C^{13}$ and $_7N^{12}$.

9. A $_{10}Ne^{20}$ nucleus has a mass of 19.99244 u. Determine the binding energy per nucleon for Ne.
10. Calculate the binding energy per nucleon in MeV for sulphur 32 that consists of 16 protons, 16 neutrons and 16 electrons.
11. What is the half-life of a radioactive nuclide whose decay constant is 2.5×10^{-8} s^{-1}?
12. What is the decay constant of $_{86}Rn^{222}$ that has a half-life of 3.82 days?
13. Compute the decay constant of $_{92}U^{238}$ that has a half-life of 4.49×10^9 years?
14. Is the following reaction possible?

$$_0n^1 + _{94}Pu^{239} \rightarrow _{53}I^{124} + _{41}Nb^{93} + 3\,_0n^1$$

15. Write the equation for U^{239} after it undergoes beta decay. Name the resulting isotope formed in the reaction.
16. In a nuclear reaction energy equivalent to 3×10^{-13} kg is released. How much is this energy in joules?
17. The fusion of deuterium is given by

$$_1H^2 + _1H^2 \rightarrow _2H^4$$

If the atomic mass of $_1H^2 = 2.0141$ u and atomic mass of $_2H^4 = 4.0026$ u, how much energy is released by the fusion of 0.5 kg of deuterium? (Use $E = (\Delta m)c^2$ and $\Delta m/m$.
18. A calcium ion has 20 protons, 20 neutrons and 18 electrons. What is the magnitude of its charge in coulombs?
19. How many photons must be absorbed by a 0.5 kg piece of meat so that the absorbed dose is 2500 Gy?
20. Baryons are composed of
   a) 2 quarks   b) 3 quarks
   c) 4 quarks   d) none of these

Appendix 1: Some Units and Abbreviations

ampere	A	Kelvin	k
atmosphere	atm	light-year	ly
British thermal unit	Btu	litre	L
calorie (physical)	cal	Metre	m
calorie (nutritional)	Cal	mile	mi
coulomb	C	minute	min
day	d	mole	mol
degree Celsius	°C	newton	N
degree Fahrenheit	°F	ohm	Ω
electron-volt	ev	pascal	pa
farad	F	pound	lb
foot	ft	radian	rad
gauss	G	revolution	rev
gram	g	second	s
henry	H	tesla	T
hertz	Hz	Unified atomic mass unit	u
horsepower	hp	volt	V
hour	h	watt	W
inch	in.	weber	Wb
joule	J	year	y

Appendix 2: Table of Isotopes

The Value listed are based on $_6C^{12}$ = 12 amu exactly. Electron masses are included.

Atomic Number Z	SYMBOL	Average Atomic Mass	Element	Mass Number A	Relative Abundance %	Mass of Isotope
1.	H	1.00797	Hydrogen	1	99.985	1.007825
				2	0.015	2.014102
2.	He	4.0026	Helium	3	0.00015	3.016030
				4	100 -	4.002604
3.	Li	6.939	Lithium	6	7.52	6.015126
				7	92.48	7.016005
4.	Be	9.0122	Beryllium	9	100 -	9.012186
5.	B	10.811	Boron	10	18.7	10.012939
				11	81.3	11.009305
6.	C	12.01115	Carbon	12	98.892	12.0000000
				13	1.108	13.003354
7.	N	14.0067	Nitrogen	14	99.635	14.003074
				15	0.365	15.000108
8.	O	15.9994	Oxygen	16	99.759	15.994915
				17	0.037	16.999133
				18	0.204	17.999160
9.	F	18.9984	Fluorine	19	100	18.998405
10.	Ne	20.183	Neon	20	90.92	19.992440
				22	8.82	21.991384
11.	Na	22.9898	Sodium	23	100 -	22.989773
12.	Mg	24.312	Magnesium	24	78.60	23.985054
13.	Al	26.9815	Aluminium	27	100	26.981535
14.	Si	28.086	Silicon	28	92.27	27.976927
				30	3.05	29.973761
15.	P	30.9738	Phosphorus	31	100	30.973763
16.	S	32.064	Sulphur	32	95.018	31.972074
17.	Cl	35.453	Chlorine	35	75.4	34.968854
				37	24.6	36.965896
18.	Ar	39.948	Argon	40	996	39.962384
19.	K	39.102	Potassium	39	93.08	38.963714
20.	Ca	40.08	Calcium	40	96.97	39.962589
21.	Sc	44.956	Scandium	45	100	44.955919
22.	Ti	47.90	Titanium	48	73.45	47.947948
23.	V	50.942	Vanadium	51	99.76	50.943978
24.	Cr	51.966	Chromium	52	83.76	51.940514
25.	Mn	54.9380	Manganese	55	100	54.938054
26.	Fe	55.847	Iron	56	91.68	55.934932

Atomic Number Z	SYMBOL	Average Atomic Mass	Element	Mass Number A	Relative Abundance %	Mass of Isotope
27.	Co	58.9332	Cobalt	59	100	57.93319
28.	Ni	58.71	Nickel	58	67.7	57.93534
				60	26.23	59.93032
29.	Cu	63.54	Copper	63	69.1	62.92959
30.	Zn	65.37	Zinc	64	48.89	63.92914
31.	Ga	69.72	Gallium	69	60.2	68.92568
32.	Ge	72.59	Germanium	74	36.74	73.92115
33.	As	74.9216	Arsenic	75	100	74.92158
34.	Se	78.96	Selenium	80	49.82	79.91651
35.	Br	79.909	Bromine	79	50.52	78.91835
36.	Kr	83.30	Krypton	84	56.90	83.91150
37.	Rb	85.47	Rubidium	85	72.15	84.91171
38.	Sr	87.62	Strontium	88	82.56	87.90561
39.	Y	88.905	Yttrium	89	100	88.90543
40.	Zr	91.22	Zirconium	90	51.46	89.90432
41.	Nb	92.906	Niobium	93	100	92.90602
42.	Mo	95.94	Molybdenum	98	23.75	97.90551
43.	Tc	*	Technetium	98		97.90730
44.	Ru	101.07	Ruthenium	102	31.3	101.90372
45.	Rh	102.905	Rhodium	103	100	102.90480
46.	Pd	106.4	Palladium	106	27.2	105.90320
47.	Ag	107.870	Silver	107	51.35	106.90497
48.	Cd	112.40	Cadmium	114	28.8	113.90357
49.	In	114.82	Indium	115	95.7	114.90407
50.	Sn	118.69	Tin	120	32.97	119.90213
51.	Sb	121.75	Antimony	121	57.25	120.90375
52.	Te	127.60	Tellurium	130	34.49	129.90670
53.	I	126.9044	Iodine	127	100	126.90435
54.	Xe	131.30	Xenon	132	26.89	131.90416
55.	Cs	132.905	Cesium	133	100	132.90509
56.	Ba	137.34	Barium	138	71.66	137.90501
57.	La	138.91	Lanthanum	139	99.911	138.90606
58.	Ce	140.12	Cerium	140	88.48	139.90528
59.	Pr	140.907	Praseodymium	141	100	140.90739
60.	Nd	144.24	Neodymium	144	23.85	143.90998
61.	Pm	*	Promethium	145		144.91231
62.	Sm	150.35	Samarium	152	26.63	151.91949
63.	Eu	151.96	Europium	153	52.23	152.92086
64.	Gd	157.25	Gadolinium	158	24.87	157.92410
65.	Tb	158.924	Terbium	159	100	158.92495
66.	Dy	162.50	Dysprosium	164	28.18	163.92883
67.	Ho	164.930	Holmium	165	100	164.93030
68.	Er	167.26	Erbium	166	33.41	165.93040
69.	Tm	168.934	Thulium	169	100	168.93435
70.	Yb	173.04	Ytterbium	174	31.84	173.93902
71.	Lu	174.97	Lutetium	175	97.40	174.94089
72.	Hf	178.49	Hafnium	180	35.44	179.94681

Atomic Number Z	SYMBOL	Average Atomic Mass	Element	Mass Number A	Relative Abundance %	Mass of Isotope
73.	Ta	180.948	Tantalum	181	100	180.94798
74.	W	183.85	Tungsten	184	30.6	183.95099
75.	Re	186.2	Rhenium	187	62.93	186.95596
76.	Os	190.2	Osmium	192	41.0	191.96141
77.	Ir	192.2	Iridium	193	61.5	192.96328
78.	Pt	195.09	Platinum	195	33.7	194.96482
79.	Au	196.967	Gold	197	100	196.96655
80.	Hg	200.59	Mercury	202	29.80	201.97063
81.	Tl	204.37	Thallium	205	70.50	204.97446
82.	Pb	207.19	Lead	208	52.3	207.97664
83.	Bi	208.980	Bismuth	209	100	208.98042
84.	Po	[210]	Polonium	210		209.98287
85.	At	*	Astatine	211		210.98750
86.	Rn	*	Radon	211		210.99060
87.	Fr	*	Francium	221		221.01418
88.	Ra	[226]	Radium	226		226.02536
89.	Ac	*	Actinium	225		225.02314
90.	Th	[232.038]	Thorium	232	100	232.03821
91.	Pa	[231]	Protactinium	231		231.03594
92.	U	[238.03]	Uranium	233		233.03950
				235	0.715	235.04393
				238	99.28	238.05076
93.	Np	*	Neptunium	239		239.05294
94.	Pu	*	Plutonium	239		239.05216
95.	Am	*	Americium	243		243.06138
96.	Cm	*	Curium	245		245.06534
97.	Bk	*	Berkelium	248		248.070305
98.	Cf	*	Californium	249		249.07470
99.	Es	*	Einsteinium	254		254.08811
100.	Fm	*	Fermium	252		252.08265
101.	Md	*	Mendelevium	255		255.09057
102.	No	*	Nobelium	254		254

Appendix 3:

1932   Werner Heisenberg* (Germany), for the development of quantum mechanics.

1935   Sir James Chadwick* (Great Britain), for the discovery of the neutron.

1937   Clinton J. Davission* (U.S.) and Sir George P. Thomson (Great Britain), for the discovery of the diffraction of electrons by crystals.

1938   Enrico Fermi* (Italy), for identification of new radioactive elements and the discovery of nuclear reactions effected by slow neutrons.

1948   Lord Patrick Blackett (Great Britain), for discoveries in nuclear physics and cosmic radiation using an improved Wilson cloud chamber.

1953   Frits Zernike (Netherlands), for the development of the phase contrast microscope.

1969   Murray Gell-Mann* (U.S.), for the study of subatomic particles.

1979   Steven Weinberg,* Sheldon Glashow* (both U.S.), and Abdus Salam* (Pakistan), for developing the theory that the electromagnetic force and the weak force are facets of the same phenomenon.

Answer Key: Answers to Odd-numbered Problems

Chapter 1

1. 3.15×10^{15}
3. b) V/m
5. b) 25×10^{-5} N
7. 3.169
9. 5.26 μC
11. 5.76×10^{-2} N/C
13. 3.77 N
15. 536.7×10^3 N/C
17. 0.084 m^2
19. 1.167 μF
21. 4.79 μF
23. 98 km/s
25. 15 ms

Chapter 2

1. d)
3. 750 mA, 8×10^{-6} Ω
5. 210 C
7. .058 mm
9. 30 Ω, 4.71×10^{-2} Ω.m
11. 189 Ω
13. 120 Ω
15. 46.09 Ω
17. 3 A, -2 A, -1 A
19. 9.054×10^{-5} V
21. 3.2×10^8 J
23. 36
25. 3.3 V
27. Rs 5.04
29. 5.086 A
31. Band 1: Blue
    Band 2: Green
    Band 3: Silver
    Band 4: Gold
33. 0.16 V – 32 V

Chapter 3

1. 1×10^{-4} T
3. 2.57×10^{-12} N, 1.54×10^{15} ms2
5. 4.8 mV
7. 1.18 Nm
9. 0.208 m
11. 0.245 A
13. 1.67×10^{-7} T
15. 51.05°
17. 5×10^{-2} Wb/A.m
19. 400 Ω
21. 0.303 Ω

Chapter 4

1. a) 50 Hz
3. 2.12 Wb
5. 6 mV
7. 320 mA
9. 50 mH
11. 1.49 V
13. 1.02 μJ
15. 4.91×10^{-5} H, 0.82 Ω
17. Step down transformer, 225 turns
19. 18.84 mV

Chapter 5

1. c) $\sqrt{2}$ V$_{rms}$
3. 1.41 A
5. a) 21.23 Ω
   b) 0.71 A
7. 9.42; 15.92 A
9. 72.11 Ω
11. 500 Ω
13. 8 W
15. 181.79 V
17. 70.7 Ω

19. 2.05×10^6 Hz
21. 50.6 nF – 49.5 nF
23. 1.875×10^2 m - 5.56×10^2 m
25. 5×10^{-3} cm

Chapter 6

1. b)
3. c)
5. b)
7. 2×10^6 Pa
9. 2.75×10^8 Pa
11. 4.0 mm
13. 1.55×10^{-4} m^2
15. $- 6.4 \times 10^{-4}$ m^3

Chapter 7

1. 0.15 mA; 15.15 mA
3. 10^5
5. 6.67; 1.0 V
7. a) Inverting amplifier
    b) Comparator

9.

A	B	Y
0	0	1
1	0	1
0	1	1
1	1	0

11. Y = 100101
13. a) all inputs are zero
15. Zero

Chapter 8

1. 6.077 μs
3. 0.483 μm
5. 1009 K
7. 1.68 eV
    505 nm
9. 65.4 μm

11. $\theta = 67.49°$
13. 2.65×10^{-19} J
    8.84×19^{-28} kg m/s
15. 2.76×10^{-36} kg
17. 4.52×10^{19} photons
19. 4.68×10^{14} Hz
    3.1×10^{-19} J
21. $2.76 \times 10\text{-}36$ kg
23. 4.14×10^4 V
    1.2×10^8 m/s
25. 1.76×10^{-33} kg m/s

Chapter 9

1. c) 3
3. 2.12 A°
5. 8.32×10^{-11} m
7. 93.7 μm
9. 364.6 nm
11. 1.88 μm
13. 1.75×10^4 V
15. i) 2.91×10^{15} Hz
    ii) 0.61×10^{15} Hz
    Frequency i) is greater than ii)

Chapter 10

1. neutron; neutral particle
3. 0.2234 u, 208.095 MeV
5. 0.009105 u
7. 0.03038 u, 28.29878 MeV, 7.0746 MeV
9. 8.032 MeV or 1.285×10^{-12} J
11. 2.772×10^7 s
13. 4.88×10^{-18} s^{-1}
15. $_{92}U^{239} + _{-1}e^o + _{93}Np^{239}$
17. 2.86×10^{14} J
19. 7.81×10^{16} photons

Index